Skywatch

Eyes-on Activities for Getting to Know the Stars, Planets & Galaxies

Peter Lancaster-Brown

Edited by Claire Bazinet

Library of Congress Cataloging-in-Publication Data
Lancaster-Brown, Peter [date]
Skywatch : eyes-on activities for getting to know the stars, planets & galaxies / Peter Lancaster-Brown.
p. cm.
Includes index.
Summary: A practical introduction to the night sky and what can be seen there with the naked eye, binoculars, and telescopes, including stars, planets, constellations, and asteroids.
ISBN 0-8069-8627-1
1. Astronomy—Observers' manuals—Juvenile literature.
[1. Astronomy—Observers' manuals.] I. Title.
QB64.L34 1992
523—dc20 92-40580
CIP
AC

Acknowledgments and credits

The author and publishers gratefully acknowledge and credit the following sources of photographic and illustrative material: American Meteorite Laboratory, pages 107, 116; Aries Press archives, plate 19, pages 10, 14, 20, 21, 22 (top), 27, 44 (top), 47, 54, 59, 63, 93, 108, 110; Australian Information Service, plate 18, page 112; Author, plates 1, 2, 3, 4, 5, 6, 8, 9, pages 9, 15, 32, 34, 37, 38, 39, 40, 41, 44 (bottom), 45, 57 (chart), 67, 76, 79, 83, 85, 86, 87, 90, 97, 114, 121; Carl Zeiss, Jena, pages 57 (bottom), 74; Mount Wilson and Palomar Observatories, plates 7, 10, 11, 12, 13, 15, 21, 22, 23; Royal Astronomical Society, pages 22, (bottom), 48, 61, 77, 81; United States Naval Observatory, plates 14, 16, 17, 20.

The author is especially grateful to his wife, Johanne, for her sometime wise counsel and her unflagging editorial assistance.

10 9 8 7 6 5 4 3 2 1

First paperback edition published in 1994 by
Sterling Publishing Company, Inc.
387 Park Avenue South, New York, N.Y. 10016

Distributed in Canada by Sterling Publishing
℅ Canadian Manda Group, P.O. Box 920, Station U
Toronto, Ontario, Canada M8Z 5P9
Distributed in Great Britain and Europe by Cassell PLC
Villiers House, 41/47 Strand, London WC2N 5JE, England
Distributed in Australia by Capricorn Link (Australia) Pty Ltd.
P.O. Box 6651, Baulkham Hills, Business Centre, NSW 2153, Australia
Manufactured in the United States of America

Sterling ISBN 0-8069-8627-1 Trade
0-8069-8628-X Paper

Contents

Introduction

You can skywatch from anywhere. Yes, I mean *anywhere.*

I have watched the stars from the open window of an apartment block. I have watched them from a remote, ice-covered volcanic island in Antarctica; the jungles of India; the pine forests of Siberia; the deserts of Western Australia. I have even watched them through the smoggy air of Central Park, right in the heart of the city of New York.

True, I did not see as many stars from Central Park as I saw from Western Australia. I did not expect to because these days the skies above New York are polluted by dust and smoke, and the bright street lights create a sky glare as in cities elsewhere. Even so, I recognized many old friends—stars whose names and patterns I had first learned before I was a teenager.

Once you have read this book, the starlit night will be a more familiar place to you.

For those who wish to make deeper explorations, using binoculars and telescopes, skywatching will provide you with a very personal eyes-on experience of the nearer universe.

But before we start to look at the sky, a cautionary note to all. As in every skill learned in life, practice makes perfect. As a beginner, do not expect to become an expert skywatcher overnight. Be practical. Many, having browsed through some popular astronomy books, expect to see stars, planets, and galaxies *exactly* as they appear in colorful pictures, *as are also seen in this book.*

Do not be misled by these pictures. Most of them were taken by the world's largest telescopes and are very much exaggerated from what you see with binoculars and small telescopes. These illustrations are put into books to catch the reader's interest and, to be frank, to pretty up the pages a little. I make no excuses for including some here because I too enjoy looking at them.

What the ordinary skywatcher actually sees of a star cluster, when he or she looks through a pair of binoculars or a small telescope, is a faint, misty patch of whitish light, often at the limits of visibility. What is more, a beginner, *at first glance,* may not see anything at all!

While photographs do arouse interest, *I* prefer the more *natural* views of a faint nebula seen through a pair of binoculars or small telescope. You will too. There can be real satisfaction in searching out and then identifying a faint distant galaxy *by your own efforts* rather than looking at it indoors in a book.

No pictures in books can measure up to being outdoors on a dark, starry night and gazing up at the countless shimmering points of light, like those in the Pleiades or in the Double Cluster in Perseus. No book can give you that firsthand thrill you get when you spot a planet through your *own* telescope.

I shall never forget my very first view of Jupiter through a newly purchased 2-inch telescope like the one shown in later pages. I was a schoolboy and had saved up my allowance for two years to buy that telescope. In the meantime I had gazed longingly in books at pictures of Jupiter taken with the world's largest telescopes. Yet, none of them could compare with my first *direct* view of Jupiter showing itself as a small, yellowish disc with two distinct light-brownish cloud belts and four tiny moons strung out alongside, glistening like pearls on a string. The view was breathtaking.

Soon after, I turned my telescope on Saturn and discovered the equally breathtaking beauty of its ring system. Although in later years I observed both planets with much larger telescopes, I have never forgotten those very first views. Even today I take particular delight in turning a small telescope on both planets to recapture some of the awe I felt on those earlier occasions.

Because I had to save for two years to buy my first telescope, I remained a naked-eye skywatcher during that period. It was time well spent, for I learned the constellations by heart and was soon able to identify all the brighter stars at a glance. To those of you who do not possess binoculars or a telescope at the moment, do not despair. One day you will possess them to be sure. In the meantime, use the opportunity to find your way about the heavens as the ancients did before telescopes were invented. Remember that skywatching can be rewarding at any level and that great discoveries have been made with *only* the naked eye even in recent times.

Peter Lancaster-Brown
Aldeburgh, 1993

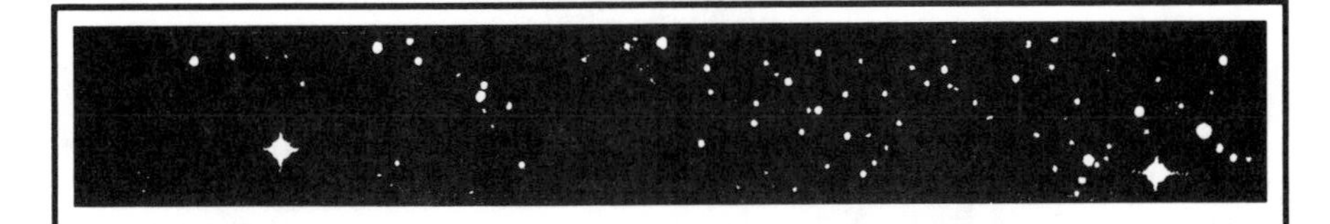

The Night Sky

Starting to Skywatch

Before going out to look at the night sky, younger skywatchers should always check with a parent or guardian first. It's important that responsible adults know exactly where you are.

Always remember to dress up warmly. Even on summer evenings it can be chilly outdoors.

When you are ready, you can start to skywatch on any clear, moonless night throughout the year. What day or month you choose does not matter, for there is always something to see.

Some stars, called circumpolar stars, will always be visible in the night sky from where you are. Other stars you will only see at certain times of the year, "in their seasons." The star maps give you this information. There are other stars you will never be able to see because they belong to the other hemisphere and are always down below your horizon.

Skywatchers who live on or near the equator are the exception. They will see no stars making circles in the sky, but *all* of the stars "in their seasons."

Skywatchers who live in the temperate (middle latitude) regions will have to remember that in summer the stars are not visible until late evening. However, the situation is different during the winter months when the stars come out in the late afternoon.

Eyes on the Stars

As this is an eyes-on book, I want you to go outdoors and start looking at the stars as soon as possible. Later I will explain things like star brightness, star colors, and the various families of stars, plus the other celestial objects like our Moon, the planets, and distant galaxies.

Later, too, we will be looking at our own home star Sol, which is the Sun, for our Sun is a star like those we see all around us in the night sky.

The only difference between our Sun, Sol, and the billions of other stars is it is much closer to Earth, our home planet. When Sol shines in the sky, its light is so powerful it blots out the more distant suns so we cannot see them during daylight.

Binoculars and Small Telescopes

If you are lucky enough to own, or able to borrow, a pair of binoculars or a small telescope, *leave them indoors for the present.*

It is much easier to learn the basic star patterns using just your own eyes. Later you *will* be using binoculars and small telescopes like those in the illustrations to look at the less bright deep-sky objects such as star clusters and galaxies.

Introducing the Night Sky

When we gaze up at the heavens on a cloudless night, the sky appears as a large dome or an inverted bowl with glittering stars sprinkled overhead in different patterns.

In cities and towns, or if the Moon is visible, you will see only the brightest stars. So long as the sky glare is not too strong, this effect, which blots out the fainter stars, can sometimes make it easier to trace the chief star patterns called constellations.

The Stars and You

On the first clear, moonless night go outdoors, or look through an open window, and introduce yourself to the night sky. For the moment forget the star maps. Just look up at the stars.

They are not alien bodies. The stars are distant cousins of ours—they are family. You do not believe it? Well, it is true. They are not alive like we are, but we have things in common. The chemicals in our bodies were all cooked in the nuclear furnaces of long-dead stars that exploded and scattered their ashes in space.

Later these ashes formed new stars *and* planets, and on the planet Earth the ashes gave rise to life. If none of the ancient stars had exploded, we would not be alive now here on Earth. We are truly star stuff—children of the universe.

Step outdoors and look up at the night sky.

Numbers of Stars We Can See

On clear, moonless nights in country areas with no street lights, we can see about 3,000 stars with the naked eye. If we use binoculars and telescopes, the number of stars we can see increases enormously. In large telescopes, you see so many it is impossible to count them all.

Seeing in the Dark

Like those of many creatures, our eyes see things better in daylight. When we go outdoors at night, we have to allow time for our eyes to get used to the darkness—or *adapt,* as we say.

Our eyes do this for us automatically. The pupil, the middle round part of the eye, grows bigger (dilates) to collect more light. The pupils of our eyes are like tiny telescopes with lenses that open up in the dark. The larger the lens in a telescope, the fainter the stars it is able to detect.

At first, when you go outdoors at night, you can barely see anything. Then slowly, as your pupils expand, objects and stars become much clearer. It is best to allow from three to five minutes for your eyes to adapt. Full adaptation takes longer. Later, when you use binoculars and small telescopes to search out deep-sky objects, you will find how important it is to allow your eyes to fully adapt.

Outdoors, when you use the star maps and lists of deep-sky objects, you will need a light to read by. An ordinary flashlight will do. Trouble is, the yellowish light from an ordinary flashlight or small lamp will dazzle your eyes and make your pupils grow smaller again. You can overcome this problem by fastening some red cellophane paper or thin red cloth over the head of the flashlight or lamp. Red light will not affect your eyes.

Star and Constellation Names

Almost all the brighter stars and all the constellations have special names. These were given to them by the ancient skywatchers who first mapped them. For example, the brightest star in the night sky is called Sirius—a name meaning "sparkling" or "burning." Sirius lies in the constellation known as Canis Major, the Larger or Greater Dog. The brighter stars in each constellation are also identified by the letters of the Greek alphabet (see box). According to this system, Sirius is the Alpha (α) star of Canis Major.

Once, at a lecture, I was asked by a budding skywatcher: "How do we know the names of the stars and planets?"

A good question, and not as silly as some might suppose.

Names, including those of the stars and planets, are given by people. In the case of the brighter stars and planets, their names were given to them by people who lived long ago. Astronomers honor tradition, and today most of the constellations, brighter stars, and planets still carry the same names as those used by the ancient Greeks and Romans, or by the Arabs who lived later.

From time to time, however, people have suggested renaming them. One astronomer tried to change the names of the constellations to biblical ones. The French once proposed changing the name of Orion's Belt to honor Napoleon. On hearing this, the English proposed Nelson instead!

In special circumstances some stars have been called after astronomers who lived in more modern times. For example, Tycho's star in Cassiopeia was an "exploding" star that Tycho Brahe observed in 1572. Another example is Barnard's star, named after the famous American skywatcher E. E. Barnard who discovered this faint star which moves faster than most others.

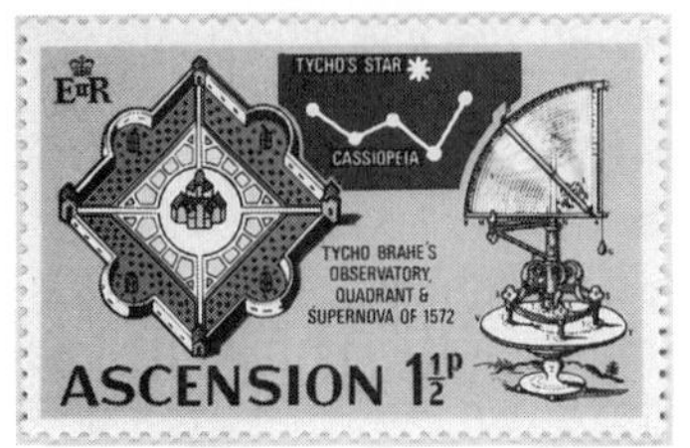

Many countries issue stamps on the themes of space travel and astronomy. This stamp, from Ascension Island, shows the position of Tycho Brahe's "new" star in 1572. You will find the position where this new star appeared on Star Map 1.

Of course, you can call the stars by your own pet names if you want to. Nevertheless, to get to know the stars properly you will *first* have to learn the traditional names printed on the star maps.

Star Designations

The principal stars in each constellation are designated by the letters of the Greek alphabet:

α	Alpha	η	Eta	ν	Nu	τ	Tau
β	Beta	θ	Theta	ξ	Xi	υ	Upsilon
γ	Gamma	ι	Iota	ο	Omicron	φ	Phi
δ	Delta	κ	Kappa	π	Pi	χ	Chi
ε	Epsilon	λ	Lambda	ρ	Rho	ψ	Psi
ζ	Zeta	μ	Mu	σ	Sigma	ω	Omega

Distances of Stars and Galaxies

While all the stars look to be at equal distances from us, this is not so. Some are much farther away than others.

For example, if we could look side on at the stars which form the constellation called Cassiopeia, we would see that those which make up its "W" shape are actually separated by huge distances.

Sometimes it is difficult for us to imagine the vast distances in space. Not counting Sol, our own star, even the nearest stars are far away by the scale of distances we use on Earth.

Astronomers know that starlight travels at 186,000 miles *every second.* Even travelling so fast, it still takes light from the nearest star over four years to reach us. This sounds even farther if we calculate that it would take a passenger jet 5 million years or an automobile 50 million years to make the same journey. And this is to the *nearest* star! Yet as astronomical distances go, it is close by.

For light to reach us from the galaxies takes several *hundred thousand* or *millions* of years.

Because distances in interstellar space (star space) are so great, astronomers find it easier to measure them by using a unit based on the speed of light. Instead of writing down the distance of the nearest star, Proxima Centauri, as 25,200,000,000,000 miles, they write more simply 4.3 light-years, which is the time it takes for light from that star to reach us.

Imagining the Solar System

Because no drawing in a book can show the real scale of space, we have to imagine what it looks like by using familiar objects and distances.

If we imagine the Sun to be the size of a basketball or a soccer ball, then the Earth will be a large pinhead at a distance of about 100 *feet* away. The Moon on the same scale will be a smaller pinhead circling the Earth at a distance of 3½ *inches.*

To reach Pluto, the farthest known planet in the solar system, we would need to move out to a distance of 1,400 *yards* from the ball. However, to reach the nearest star, Proxima Centauri, we would need to travel 5,500 *miles.* To reach the nearest galaxy, it would mean travelling around the world over 7,000 *times.*

The Distances of the Sun and the Planets

In contrast to the vast distances between the other stars, Sol, our own star, lies only 8 minutes light distance away, or 93 million miles.

The distances between Sol's family of planets that form the solar system are much smaller compared with the distances between stars so they are measured in miles (or kilometres). For example, the Moon is less than one-and-a-half light *seconds* distant from the Earth, or about 240,000 miles. The farthest planet, Pluto, is approximately 5½ light *hours* from the Sun, or 3,666 million miles. By astronomical standards, the solar system is our own backyard in space.

How and Why the Stars Appear to Move

When you watch the stars carefully on a clear, moonless night, you will soon notice that they are not fixed in one spot but slowly shift around the sky. You will also see that it is the whole star sphere which is shifting, not the separate stars.

It is easier to spot this slow movement if you line up a bright star, or a group of stars, against a tall building, tree, or pole and then stay put to watch. Five minutes will be long enough to reveal a shift.

If you look at the sky for a longer period, you will notice that certain stars rise in the east, climb higher, then gradually sink down and set over the western horizon.

If you had time to watch over an even longer period, you would notice that the stars which rose in the east, and later set in the west, reappear in the east in about 24 hours.

While it is the stars that *appear* to make an ever-repeating circuit around the sky, they are in fact not moving at all. *Instead it is the Earth that is spinning on its axis in about 24 hours,* and this is what gives the impression it is the whole star sphere itself that is shifting.

Finding the Distances of Stars

Astronomers find the distances of the nearer stars by using a method similar to the familiar game of holding a hand over one eye at a time, or winking, to watch objects in the foreground jump from side to side.

Astronomers call the process finding the parallax of a star.

To do it, astronomers use the diameter of the Earth's orbit. A starfield is photographed when the Earth is on one side of its orbit, then photographed again six months later when the Earth is on the other side.

When the two photographs are compared, the nearest (foreground) stars jump slightly. By measuring the amount of the jump, or parallax, and taking into account the size of the Earth's orbit, astronomers can calculate how far away the stars are.

However, the stars are so far away that the jump is tiny. Even the nearest star jumps *less than one quarter of the width of a human hair.*

Circumpolar Stars

Unless you live on the equator or near it, which most of you do not, some of the stars you see in your sky are always visible every clear night. These stars never rise and set. Instead they appear to circle the sky, taking the same 24-hour period to do so as those stars which do rise and set. The stars which never rise and set are called the circumpolar stars, for they continually circle the north and south poles.

The northern circumpolar stars *appear* to circle the night sky. For this photograph, a camera was set for a long exposure, over several hours, so that the star images made trails as the Earth rotated. The Pole star, Polaris, is the bright semi-circle near the center. If the camera could have been left for 24 hours, the trails of all circumpolar stars would have formed complete circles.

What stars and constellations are circumpolar depends on where you are living on the Earth's surface. The closer you live to either of the poles, the more stars become circumpolar. Right on the equator *no* stars are circumpolar, and all will rise and set in "their seasons."

If you live in the northern hemisphere, you will see the circumpolar stars that circle the north celestial pole. If you live in the southern hemisphere, you will see the stars that circle the south celestial pole. For most readers these stars are the ones shown in the round Star Maps 1 and 5. The north and south points in the sky around which the circumpolar stars circle, or turn, mark the axis line around which the Earth is spinning in about 24 hours.

Finding the Compass Points

It is easier to make yourself familiar with how the stars appear to rise and set if you know where the directions north, south, east, and west are in your area.

If you do not have a compass to find these directions, you can best find out in daylight when the Sun is shining.

At local midday, the Sun will be due south of you if you are in the northern hemisphere and due north of you if you live in the southern hemisphere.

Use a stick in the ground to cast a shadow, and at local midday the shadow will mark out your north-south line. Your east-west line then runs at right angles to this.

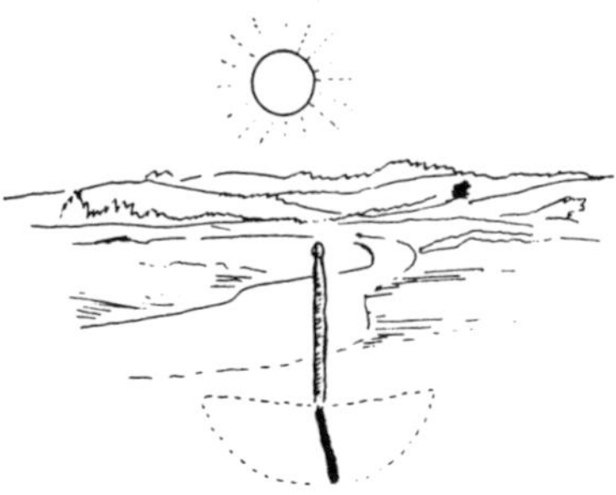

Using the Sun to find your north and south points.

Note that the Sun is not always *exactly* due south of you at midday. This is due to something known as the equation of time, but it need not bother you. The directions you will find by the Sun method are good enough to provide the four compass points approximately.

Of course, if you own or can borrow a pocket compass, you can check one method against the other.

If you want to experiment further, using your local Sun shadow, you can lean a stick so it points to the celestial pole in your hemisphere. This will give you a simple sundial to indicate the local time.

The northern hemisphere has a star called the Pole star (also called Polaris or Alruccabah) which marks the north celestial pole. The southern hemisphere is not so fortunate; no bright star marks the south celestial pole.

The Stars in Their Seasons

While the stars which are circumpolar for *you* are always visible throughout the night, the rest of the stars rise and set and come and go with the seasons. For most skywatchers, these seasonal stars are the ones shown in Star Maps 2, 3, and 4.

The Nearest Stars

Star	Apparent Magnitude	Distance (Light-Years)
Proxima Centauri	10.7	4.3
α Centauri A	0.0	4.3
α Centauri B	1.4	4.3
Barnard's Star	9.5	5.9
Wolf 359	13.7	7.6
Lalande 21185	7.5	8.1
Sirius A	–1.4	8.7
Sirius B w	8.7	8.7
Luyten 726-8 A	12.4	8.7
Ross 154	10.6	9.3
Ross 248	12.3	10.3
ε Eridani	3.7	10.7
Ross 128	11.1	10.9
Luyten 789-6	12.6	11.0
61 Cygni A	5.2	11.2
61 Cygni B	6.0	11.2
Procyon A	0.5	11.4
Procyon B w	10.8	11.4
ε Indi	4.7	11.4
Struve 2398 A	8.9	11.5

w = white dwarf star

If you select and watch a particular star as it appears to move in relation to a building, tree, or pole, as you did earlier, you will notice that it does not reappear exactly in the same spot at the same time as the previous night but about four minutes *earlier.* Over a period you will notice that the time of the star's reappearance advances four minutes *every* night.

This daily advance adds up from night to night so that after two weeks it will be 60 minutes (one hour) ahead. This means that if you want to see a star or constellation in exactly the same position as before, at the end of two weeks you will have to look out one hour (60 minutes) earlier. At the end of four weeks it will be two hours earlier and so on.

This daily advance of the whole star sphere by four minutes every night is caused by the fact that *the Earth, in addition to rotating (spinning) on its axis over 24 hours (actually 23 hours, 56 minutes, and 4 seconds), is at the same time travelling around (orbiting) the Sun at a distance of 93 million miles in a period of one year or 365¼ days.*

The annual journey of the Earth around the Sun makes the stars change their positions from night to night and from season to season. Exactly twelve months later, the Earth will have returned to the place in its orbit where it started from a year earlier, and the changing star patterns begin to repeat.

The practical consequences of this gradual shift of the whole star sphere is that the *non*-circumpolar stars and constellations you see in the middle parts of the sky are different in winter, spring, summer, and autumn at a particular time of night. The Star Maps 2, 3, and 4 are drawn up to indicate the months of the year a star or constellation is best seen in its season.

The Two Hemispheres of the Sky

When we look up at the night sky, we see only *half* the stars that are visible to the naked eye. Underneath us, lying below the horizon, is another unseen dome, or inverted bowl, covered with stars.

The other star dome is hidden from view because the Earth is in the way. However, astronauts journeying to the Moon see the other star dome, or hemisphere, as well because the Earth then does not block it out.

As the Earth gets in our way, many of the stars seen in North America are different from those seen in Australia. It is for good reason that Australia is often humorously referred to as the country "Down Under."

Star Maps and Constellations

Without star maps to guide us, the night sky looks like a very complicated and confusing puzzle. In many ways it is a bit like one of those mystery puzzles you see in magazines where you connect up a series of numbered dots to suddenly discover a hidden shape or picture.

Of course, the stars in the sky are not numbered that way. Instead we have to look at our star maps to see how the lines joining the stars have been drawn in by astronomers to form star patterns called constellations.

Long ago the early Babylonian and Greek skywatchers used human figures, animals, and familiar objects to represent the constellation patterns. They invented fascinating legends telling how these names came to be in the sky. Today we still use almost the same names for the constellations as the earlier skywatchers did about two thousand years ago.

To our modern eyes, however, some of the figures and animal shapes do not seem to fit over the background stars as well as they might. While we still retain the old names for the constellations, modern star maps do not print in the colorful shapes. In their place, modern star maps use imaginary lines to link up the stars in a constellation. These form simple geometrical patterns which help us recognize one constellation from another. Unlike the older star maps, modern star maps have definite boundary lines to mark out each constellation separately.

Number of Constellations

In the past some of the older constellations were found to be too big and sprawling, so astronomers subdivided them. The southern constellation of Argo Navis, which commemorated the legend of Jason's ship, the *Argo*, used in the search for the Golden Fleece, was divided off to form four separate modern constellations called

Carina, the Keel; Puppis, the Poop; Vela, the Sails; and Pyxis, the Compass.

Now 88 separate constellations are recognized and these cover both hemispheres. Some of the constellations in the far southern half of the sky are more modern. As the Babylonians and the Greeks could not see to those far-off regions, these constellations were invented by people who travelled there later.

Naming the Constellations

In different countries, people have different local names for the constellations. In North America, the Great Bear is often called the Big Dipper because, rather than a Bear as the Greeks imagined it, its stars suggest the shape of a dipper, ladle, or long-handled cup.

In Britain, and in some other parts of Europe, the same constellation is known as the Plough (Plow) or the Wagon, since its shape also suggests these two farming objects.

Elsewhere in Europe, it is known as the Butcher's Cleaver, while in some parts of France they call it the Saucepan or Casserole.

If we had to print in all the local names on our star maps, it would be confusing, and there would be little room left to put in the stars and other interesting objects. Instead, astronomers have agreed to use scientific names for constellations, and these are based on the old Greek names. All are in Latin, but most are quite easy to remember. Some names can be pronounced more than one way. A list of all the 88 constellations with their Latin and more common names is given below.

Signpost Constellations

Some constellations have brighter stars than others and more distinct shapes. Because these are easier to recognize, we can use them as signposts to locate less bright constellations. Fortunately, there are several "easy" constellations scattered in both hemispheres.

In the northern hemisphere, one of the easiest signpost constellations to find is Ursa Major, the Great or Big Bear. In North America, it is better known as the Big Dipper. In Europe, it is called the Plough. Other countries have different names for it.

Once Ursa Major is recognized, it can be used to point to Ursa Minor, the Little or Smaller Bear. It also points to the Pole star that indicates true north and the hub in the sky around which the northern circumpolar stars turn.

The ancient skywatchers imagined the constellation patterns formed by the stars to be in the shape of human figures, animals, and other familiar objects.

Traditional Patterns of the Northern Hemisphere

In the southern hemisphere, an easy constellation to locate is Crux, the Southern Cross. Although it was seen by the Babylonians, its name was given to it by those who came later. It is a very important signpost constellation and appears on the national flags of Australia and New Zealand. In Western Australia, the constellation was used by gold prospectors to guide them through the arid regions. They were so grateful that later they called their new township after it. The town of Southern Cross still flourishes today. It is a unique place, for all its streets have astronomical names!

Traditional Patterns of the Southern Hemisphere

The image of the Southern Cross is important in many countries of the Southern Hemisphere and appears in place names, and on flags, postage stamps, and air mail stickers. The three stamps from Botswana depict *(left to right)* Orion, Scorpius, and the Southern Cross.

The Southern Cross and the Coal Sack nebula.

The Constellations

Constellation	Genitive ending	Meaning	Abbreviation
Andromeda	-dae	Chained Maiden	And
Antlia	-liae	Air Pump	Ant
Apus	-podis	Bird of Paradise	Aps
Aquarius	-rii	Water Bearer	Aqr
Aquila	-lae	Eagle	Aql
Ara	-rae	Altar	Ara
Aries	-ietis	Ram	Ari
Auriga	-gae	Charioteer	Aur
Bootes	-tis	Herdsman	Boo
Caelum	-aeli	Chisel	Cae
Camelopardus	-di	Giraffe	Cam
Cancer	-cri	Crab	Cnc
Canes Venatici	-num -corum	Hunting Dogs	CVn
Canis Major	-is -ris	Great Dog	CMa
Canis Minor	-is -ris	Small Dog	CMi
Capricornus	-ni	Sea Goat	Cap
Carina	-nae	Keel of the Ship	Car
Cassiopeia	-peiae	Lady in Chair	Cas
Centaurus	-ri	Centaur	Cen
Cepheus	-phei	King	Cep
Cetus	-ti	Whale	Cet
Chamaeleon	-ntis	Chameleon	Cha
Circinus	-ni	Compasses	Cir
Columba	-bae	Dove	Col
Coma Berenices	-mae -cis	Berenice's Hair	Com
Corona Australis	-nae -lis	Southern Crown	CrA
Corona Borealis	-nae -lis	Northern Crown	CrB
Corvus	-vi	Crow	Crv
Crater	-eris	Cup	Crt
Crux	-ucis	Southern Cross	Cru
Cygnus	-gni	Swan	Cyg
Delphinus	-ni	Dolphin	Del

Constellation	Genitive ending	Meaning	Abbreviation
Dorado	-dus	Swordfish	Dor
Draco	-onis	Dragon	Dra
Equuleus	-lei	Little Horse	Equ
Eridanus	-ni	River Eridanus	Eri
Fornax	-acis	Furnace	For
Gemini	-norum	Heavenly Twins	Gem
Grus	-ruis	Crane	Gru
Hercules	-lis	Kneeling Giant	Her
Horologium	-gii	Clock	Hor
Hydra	-drae	Water Monster	Hya
Hydrus	-dri	Sea-Serpent	Hyi
Indus	-di	Indian	Ind
Lacerta	-tae	Lizard	Lac
Leo	-onis	Lion	Leo
Leo Minor	-onis -ris	Small Lion	LMi
Lepus	-poris	Hare	Lep
Libra	-rae	Scales	Lib
Lupus	-pi	Wolf	Lup
Lynx	-ncis	Lynx	Lyn
Lyra	-rae	Lyre	Lyr
Mensa	-sae	Table (Mountain)	Men
Microscopium	-pii	Microscope	Mic
Monoceros	-rotis	Unicorn	Mon
Musca	-cae	Fly	Mus
Norma	-mae	Square	Nor
Octans	-ntis	Octant	Oct
Ophiuchus	-chi	Serpent Bearer	Oph
Orion	-nis	Hunter	Ori
Pavo	-vonis	Peacock	Pav
Pegasus	-si	Winged Horse	Peg
Perseus	-sei	Perseus	Per
Phoenix	-nicis	Phoenix	Phe
Pictor	-ris	Painter's Easel	Pic

Constellation	Genitive ending	Meaning	Abbreviation
Pisces	-cium	Fishes	Psc
Piscis Austrinus	-is -ni	Southern Fish	PsA
Puppis	-ppis	Poop (Stern of Ship)	Pup
Pyxis (= Malus)	-xidis	Compass	Pyx
Reticulum	-li	Net	Ret
Sagitta	-tae	Arrow	Sge
Sagittarius	-rii	Archer	Sgr
Scorpius	-pii	Scorpion	Sco
Sculptor	-ris	Sculptor	Scl
Scutum	-ti	Shield	Sct
Serpens (Caput and Cauda)	-ntis	Serpent (Head and Tail)	Scr
Sextans	-ntis	Sextant	Sex
Taurus	-ri	Bull	Tau
Telescopium	-pii	Telescope	Tel
Triangulum	-li	Triangle	Tri
Triangulum Australe	-li -lis	Southern Triangle	TrA
Tucana	-nae	Toucan	Tuc
Ursa Major	-sae -ris	Great Bear	UMa
Ursa Minor	-sae -ris	Little Bear	UMi
Vela	-lorum	Sails	Vel
Virgo	-ginis	Virgin	Vir
Volans	-ntis	Flying Fish	Vol
Vulpecula	-lae	Little Fox	Vul

The Signs of the Zodiac

The ancient skywatchers noted that some stars were not fixed in place but wandered around the middle parts of the heavens. They also noticed that the wandering stars followed the same line as the annual path of the Sun. These stars they called planets—a name from the Greek meaning "wanderer." Of course, *we* know that the Sun itself does not actually move around the sky. It only appears to move because the Earth itself is moving around the Sun. The

Sun's annual movement is an illusion, but the ancient astronomers did not realize this.

The narrow zone in the sky traced out by the Sun and the wandering stars was divided by the ancient Greeks into twelve special constellations, or signs, to which they gave the name the zodiac—a Greek word meaning "circle of animals." The name is a reference to the fact that many of the Greek constellations were called after living creatures.

The actual line along which the Sun and the wanderers moved through the signs they named the ecliptic, and this line is still found on modern star maps.

The first sign, or zodiacal constellation, is Aries, the Ram, through which the Sun *appears* to travel during the month of March. The second sign is Taurus, the Bull, through which the Sun travels in April, and so on. The Sun enters a different sign (constellation) for each month of the year, traveling in turn through all twelve signs. It comes back to start again in Aries in March.

Early in history, astrology and fortune-telling went hand in hand with more scientific astronomy. The zodiacal signs were of special significance to the Greeks in forecasting future events and the fate of humans. The soothsayers, who made these forecasts, truly believed that the stars and planets had direct influence over the lives of humankind.

Predictions about what would happen to a person were based on when the Sun, Moon, or planets entered a particular sign. A person's birth sign—the zodiacal sign through which the Sun was passing at the time of their birth—and the positions of the wandering stars at that critical moment were supposed thereafter to influence a person's character, career prospects, and general day-to-day fate. Surprisingly many people still believe this today, which is why you see plenty of horoscopes printed in newspapers and magazines.

Nevertheless, there is *one* wandering star that travels the zodiac which determines all our destinies. This is, of course, Sol, our friendly neighborhood power plant. Without Sol's daily outpouring of energy to warm our planet and grow the food we eat to sustain our bodies, our days on Earth would be numbered. In this way then, at least, one wandering star does determine our future, but in a very different way to that written about by astrologers.

A zodiac from medieval times showing the twelve signs with the Sun moving in Libra. The Earth (shown as a castle) and the Moon are placed at the center because the astronomers then believed the Sun travelled around the Earth.

The Signs of the Zodiac

1 Aries, the Ram (Mar 21–Apr 20)
2 Taurus, the Bull (Apr 21–May 21)
3 Gemini, the Twins (May 22–June 21)
4 Cancer, the Crab (June 22–July 23)
5 Leo, the Lion (July 24–Aug 23)
6 Virgo, the Virgin (Aug 24–Sept 23)
7 Libra, the Scales (Sept 24–Oct 23)
8 Scorpio, the Scorpion (Oct 24–Nov 22)
9 Sagittarius, the Archer (Nov 23–Dec 22)
10 Capricorn, the Goat (Dec 23–Jan 20)
11 Aquarius, the Water-Bearer (Jan 21–Feb 19)
12 Pisces, the Fishes (Feb 20–Mar 20)

Claim Your Own Zodiacal Constellation

Without having to believe in birth signs and horoscopes, we can have a little fun and adopt our very own birth sign for special observation.

In this way, you have a constellation that will always be special just for you.

To find out which is *your* birth constellation, check the date against the list above. For example, if your birthday falls on August 24 (any year), your birth constellation is Virgo. If, on the other hand, you were born the day before, on August 23 (any year), your birth constellation is Leo.

Some of you will be "luckier" than others and have a personal constellation rich in bright stars, clusters, and galaxies. My own is Aries, which is rather dull. Nevertheless, even if you have a dull constellation, a bright planet will sometimes cross over it. To find out when this will happen, look it up in the Planet-Finder Tables.

Celestial Longitude and Latitude

When navigators go on voyages, they fix their places on the Earth's surface by using longitude and latitude. Likewise astronomers, on their celestial voyages, need to have fixed points in the sky. Instead of calling them longitude and latitude, they call them Right Ascension (RA for short) and Declination (Dec).

You can easily find your way around the night sky without bothering to learn any reference-point system. The star maps tell you when you can find a particular constellation at any time of night and any time of year. While this system works very well for skywatchers using the naked eye, binoculars, and small telescopes, it becomes very impractical for astronomers maneuvering large telescopes.

By knowing beforehand the Right Ascension and Declination of a celestial object, astronomers can set their telescopes *exactly* to the right spot. Now all such controls are done by push-button computers.

All scientific star maps show the lines of Right Ascension and Declination. Right Ascension is shown on our maps running along the bottom edge and is given in hours, minutes, and seconds, starting at zero (0) hours then running clockwise to 23 hours, 59 minutes (23h59m). Declination is shown at each side on our maps and runs + (upwards) and – (downwards). It starts on the line called the (celestial) *equator* and is given in degrees (°).

Examples of Right Ascension and Declination

Turn to Star Map 2 and seek out the famous bright star Sirius (Alpha Canis Majoris). Its approximate position is RA 6h40m and its Dec is –16°. Using the same star map, now look for the bright star Betelgeuse (sometimes nicknamed Beetle Juice!) also known as Alpha Orionis. Betelgeuse's RA is 5h55m and its Dec is +8°.

Note: While some of the tables of objects provided below include RA and Dec information, you can ignore them if you want to because you can find all our celestial objects by direct reference to the constellation. However, it is necessary for you to know how things are measured in the sky. In more advanced books you will find many objects listed by their RA and Dec.

When planet spotting, you will be using another measure called celestial longitude. This is a cousin of RA. Instead of using hours, minutes, and seconds, celestial longitudes are given in degrees and measured along the *ecliptic* from zero to 359 degrees (0° to 359°).

Star Names and Star Maps

Many of the names given to the stars are a legacy from the Arab skywatchers who followed the Greeks. Although many are corruptions of earlier names, these star names have become so traditional that it would be difficult to change them now.

Some stars have more than one name, but on our star maps only the best known are included.

Most star names are practical descriptions given by the Arabs to where stars belong in their constellations. For example, Dubhe, the upper pointer star in the signpost constellation of Ursa Major, means "the Back of the Greater Bear," and this is where you can actually see it when you look at star maps showing the old Greek constellation figures.

Likewise, Merak, the lower pointer star which lies below Dubhe, means "the Loins of the Greater Bear."

In the same constellation the star Phad, or its alternative name Phekda, means "the Thigh of the Greater Bear." Its neighbor Megrez means "the Root of the Tail."

This method of identifying a particular star, according to where it lies, could be very confusing because not everyone agreed on how the different parts of a constellation matched up over individual stars.

A more scientific and convenient method was devised in 1603 by the German mapmaker Johann Bayer. He used the letters of the Greek alphabet to label individual stars.

In Bayer's system the brightest star in a constellation is designated Alpha (α), the next brightest Beta (β), the third brightest Gamma (γ), and so on. This is the system in use today.

While it is still customary to refer to a bright star by its popular name, for example Polaris (the Pole star), it is also usual to use its more scientific designation. Since Polaris is the brightest star in the constellation of Ursa Minor, it is referred to as Alpha (α) Ursae Minoris. Note that on star maps we use the letter α rather than spelling out Alpha. It is important also to note that when you refer to a star according to Bayer's system, you have to use what is called the genitive form of the constellation name. Instead of Ursa Minor, it becomes Urs*ae* Minor*is*. This is simply a way of saying "It belongs to this constellation." The genitive form for all 88 constellations can be found on pages 23–25.

Other Star Designations

As well as having Arab names, and designations using the Greek alphabet, many stars carry additional ones invented by later astronomers.

Beginners might want to skip over this section and come back to it later because, for casual starwatching, the old Arab names and the Greek letters are sufficient. Eventually, however, when you become more skillful, you will need to know about the other ways stars (and other celestial objects) are designated.

For example, when the first British Astronomer Royal, John Flamsteed, formed his own list, or catalogue, of stars, he used numbers. In this way Aldebaran (the Arab name meaning "the follower"), the Alpha star of Taurus in Bayer's method, also becomes star number 87 in Taurus. Note that Flamsteed did *not* number the stars in order of brightness.

Later still, this same star also came to be known as 8639 L1 in a list, or catalogue, compiled by the French astronomer J. J. Lalande (who is rumored to have eaten *live spiders*, but that is another story).

Aldebaran is also known as 1420 BAC in a catalogue made for the British Association.

More advanced star maps than those in this book often use these designations, especially for fainter stars which do not have an Arab name or a Greek letter.

In addition to those for ordinary stars, there are lists and catalogues for variable stars, star clusters, galaxies, meteor showers, and comets.

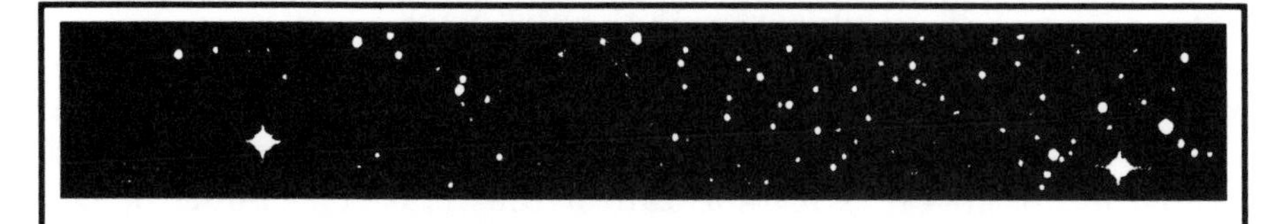

Outdoors with the Star Maps

Finding Stars and Constellations the Year Round

There are five main star maps in this book.

If you have access to a copying machine, you might, for a small sum, run off some enlargements for your *own* use outdoors. These copies will be easier to read and save wear and tear on the book.

If you live in the northern hemisphere, choose Star Map 1; for the southern hemisphere, choose Star Map 5.

Star Maps 1 and 5 are circular in shape. Except for skywatchers who live on or near the equator, one of these maps represents the circumpolar stars in *your* hemisphere. These are the stars that ever-circle your sky.

Which way up you hold the star maps depends on the month you go out to study the stars.

As long as the sky is dark, cloud-free, and the Moon is absent, you can start to skywatch any time.

Northern Hemisphere—Star Map 1

To show how the star maps work, we will take the month of September as an example.

You will need to know the different compass directions in your locality. (If you don't, turn to Finding the Compass Points on page 15 or look at a compass.)

Facing north, hold the map so that the month of September, printed along the circular edge of the map, is at the top.

The map now shows you how all the circumpolar stars are arranged in the sky at *9 p.m.* for September any year.

Now, looking up at the sky, Ursa Major (the Big Bear, or Dipper) is in the northwest, with the tail of the Bear (or handle of the Dipper) pointing westward.

Make sure you have correctly identified Ursa Major, then look at its two leading stars Dubhe (α) and Merak (β). When reading the star maps outdoors, use a red-tinted light as ordinary light will dazzle your eyes (see also Seeing in the Dark on page 9).

Dubhe and Merak are called the Pointer Stars because they point, or signpost, the North Star (Polaris).

To find Polaris (α Ursae Minoris) extend a line *upwards* by about five times the distance between Merak and Dubhe.

In addition to signposting the North Star, this also locates for you Ursa Minor, whose stars are a little fainter than those of Ursa Major.

At 9 p.m. in September, the Little Bear lies over on its back.

You have now found two signpost constellations.

A third signpost constellation is Cassiopeia, the Lady in the Chair.

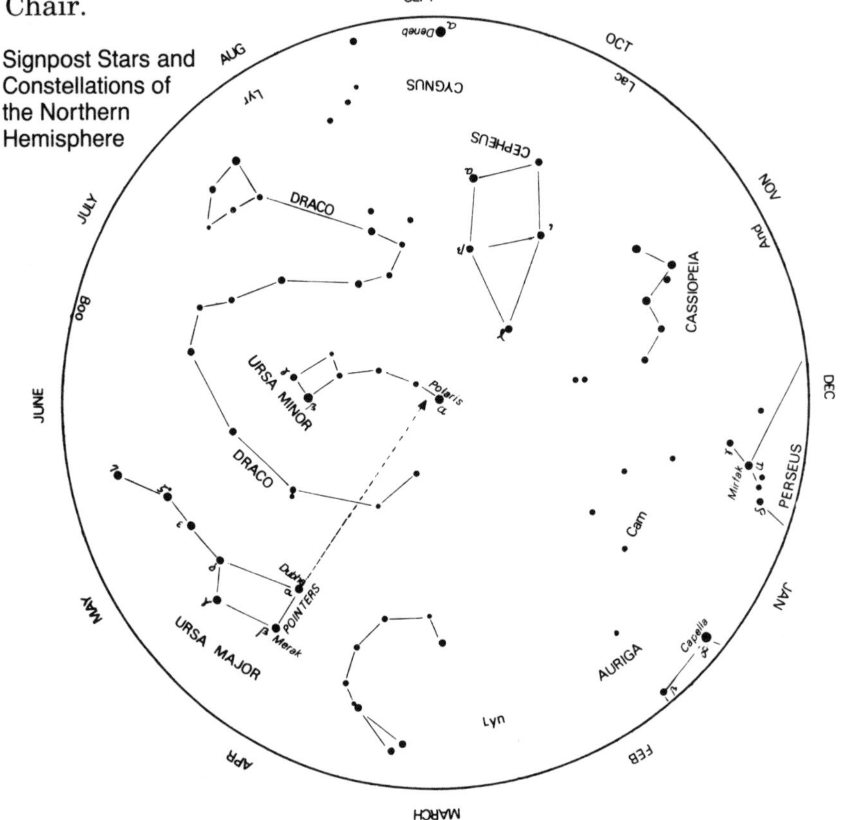

Signpost Stars and Constellations of the Northern Hemisphere

Her "W" shape can be seen tilted over on its side, high in the eastern sky.

From Ursa Major, Ursa Minor, and Cassiopeia you will be able to trace out the other, fainter circumpolar constellations such as Cepheus, the Warrior King, seen directly above your head, and Draco, the Dragon, weaving his sinuous way between Ursa Major and Ursa Minor.

What stars you see in *your* night sky will depend on how dark the sky is where you are. Street lights and misty or moonlit nights will blot out some of the fainter stars and constellations.

What if you want to look at the northern stars at 11 p.m., or at any other time of night, instead of 9 p.m. in September?

Simple.

The circular star maps are *also* a kind of adjustable 24-hour clock face.

Each month of the year is also equal to two hours in clock time.

So, to find how the stars will look at 11 p.m. in September, *turn the map around one month* so that *October* is now at the top.

This will show you how all the circumpolar stars have swung around in two hours. If you want to look at 10 p.m., you would turn the map around only *half* a month. If you want to skywatch before 9 p.m., say at 7 p.m., turn the star map *the other way* so that *August* is at the top.

This adjustable system allows you to look at the sky any time of the night, any month of the year. All you have to remember is that each month on the map is set for 9 p.m. For any other time of night in that month you have to turn the map *one month either way* for every two hours difference in observation time.

Southern Hemisphere—Star Map 5

If you have not read through the section for the northern hemisphere, do it now *indoors* while holding Map 1 in front of you pretending to skywatch. Although you will not be able to see the same circumpolar stars outdoors, it will teach you how the star maps work.

Then, to skywatch in the southern hemisphere, choose Map 5.

Facing south and, still using the month of September as our example, hold the map so that the month of September is at the top.

This shows you how all the circumpolar stars are arranged in the sky at 9 p.m. for September any year.

Looking at the sky, Crux, the Southern Cross, will be low in the southwest sky.

Trace out the shape of the Cross formed by the stars α, β, γ, and δ. Be sure you have identified the Cross properly (see below).

To find the approximate position of the south celestial pole, you can use the stars labelled α and β in the neighboring constellation of Centaurus, the Centaur.

Alpha Centauri is sometimes called Toliman, or Rigel Kentaurus, meaning the Centaur's Foot. It is the third brightest star in the night sky. Beta Centauri is often called Hadar.

These two stars are used as Pointer Stars like Dubhe and Merak in the northern hemisphere.

By projecting a perpendicular line from the Pointers and then projecting a second line along the longer axis of the Southern Cross, they will intersect at a point near the south celestial pole.

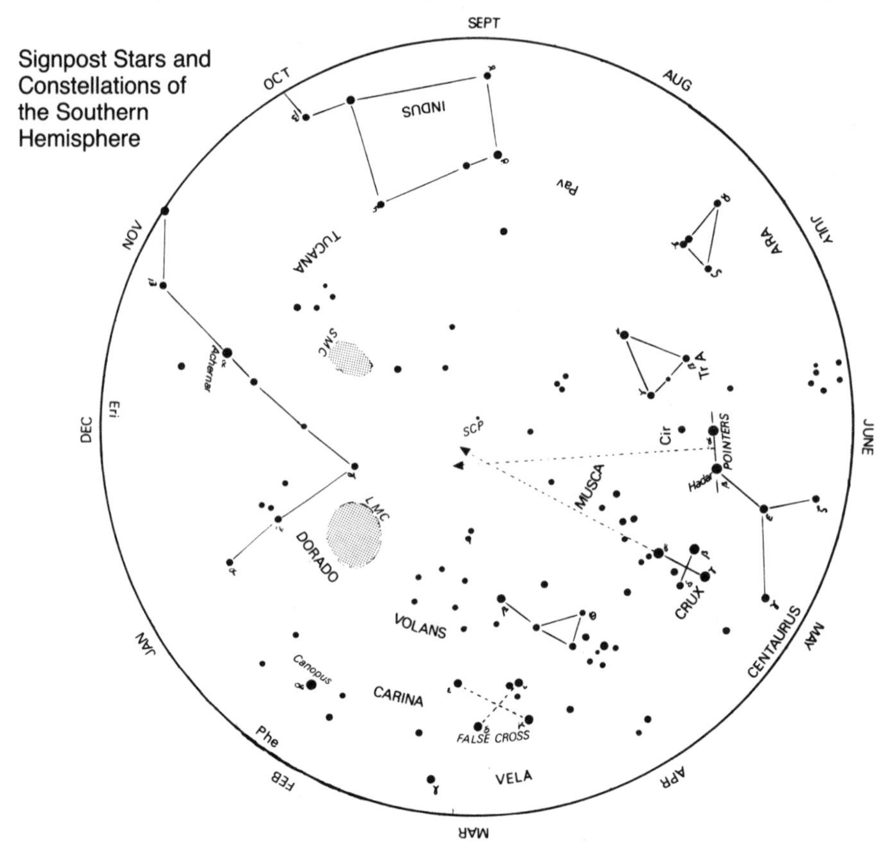

Having located the constellations of Crux and Centaurus, and the point in the sky around which the southern circumpolar stars circle, you can use them as signposts to locate other stars and constellations.

Canopus (α Carina), named after a famous ancient ship's pilot, and the second brightest star in the sky, is seen low in the east.

Achernar (α Eridani), a bright star whose name means the Tail of the River, lies to the east.

The visibility of the two Magellanic Clouds will depend on your local sky conditions. If the sky is dark and the street lights are dim, you will see them as small patches of light.

The Smaller Cloud (Nubecula Minor) lies on the other side of the south celestial pole from Crux. The Larger Cloud (Nubecula Major) is immediately below the Smaller Cloud.

If you wish to skywatch at different times of night from 9 p.m., the same rules apply as with the northern stars; that is, one calendar month equals two hours of clock time.

Crux, the Southern Cross, although usually thought of as a southern constellation, can also be seen from parts of the northern hemisphere. If you live in Florida or southern Texas, Crux will appear in its season low above the southern horizon, but it will never rise very high.

Likewise, in the southern hemisphere, Ursa Major, the Great Bear, usually considered a northern constellation, appears in its season low in the northern sky in certain parts of Australia.

NOTE: When trying to identify the Southern Cross, it is important to know that there is *another* cross of stars nearby known as the False Cross. The False Cross is formed by some bright stars belonging to the constellations of Carina, the Keel; and Vela, the Sails. If you use your star map correctly, there will be no mistaking which cross is which.

The Middle Sky—Star Maps 2, 3, and 4

Maps 2, 3, and 4 represent the stars and constellations which appear to rise and set in their seasons.

Skywatchers on or near the equator do not have any circumpolar stars in their sky. Instead, *half* of the circumpolar stars *on each of the circular Maps 1 and 5* will be visible to them at any one time. These stars are seen *as extensions on either side of the middle stars on Maps 2, 3, and 4.* These extension stars, like the middle stars, will rise and set in their seasons.

As with Maps 1 and 5, Maps 2, 3, and 4 are set for 9 p.m. for any particular month of the year.

The month of the year is printed along the *bottom* edge of the maps.

Again, let the month of September provide our working example.

For 9 p.m. in September, choose Star Map 4.

High in the sky is Cygnus, the Swan. This distinct constellation looks more like a large cross leaning over on its side. For this reason it is often called the Northern Cross. Its brightest star is Deneb (α Cygni).

To the west of Cygnus is Lyra, the Harp. Its brightest star is Vega (α Lyrae), the fifth brightest star in the sky.

Below Cygnus is Aquila, the Eagle. Its leading star, Altair (α Aquilae), is the twelfth brightest star in the sky.

Notice that, together, Deneb, Vega, and Altair form a large triangle. This triangle of stars provides a signpost to help you fix the fainter constellations.

Note that the whole star sphere appears to shift slowly westwards (owing to the *Earth's* rotation in 24 hours), and therefore at different times of the night you see different stars cross your north–south line (called the meridian). Stars will rise in the east, climb higher and cross your meridian, then sink lower and finally set in the west.

Like the circular maps, the rectangular maps can be adapted to suit viewing the stars at different times of the night from 9 p.m. by going backwards or forwards one month for every two hours clock-time difference.

By 11 p.m. in September, Cygnus, Aquila, and Lyra will have shifted westwards but will still be visible. Since 9 p.m., other stars will have risen in the east.

By 1 a.m. in September, the constellation of Pegasus, the Winged Horse, which was still low in the east at 9 p.m., will now be near the meridian.

The Northern Circumpolar Stars

Star Map 1

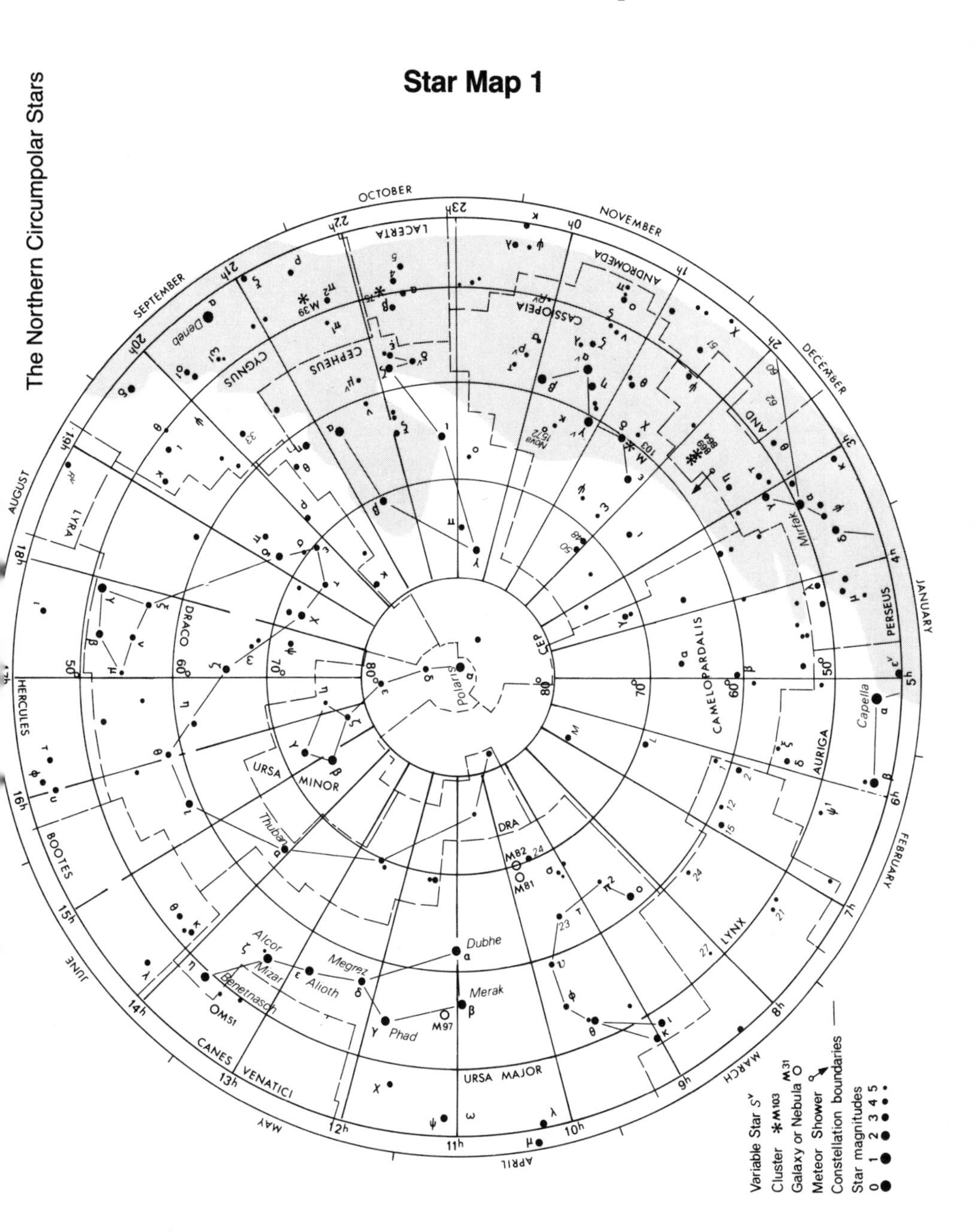

Star Map 2

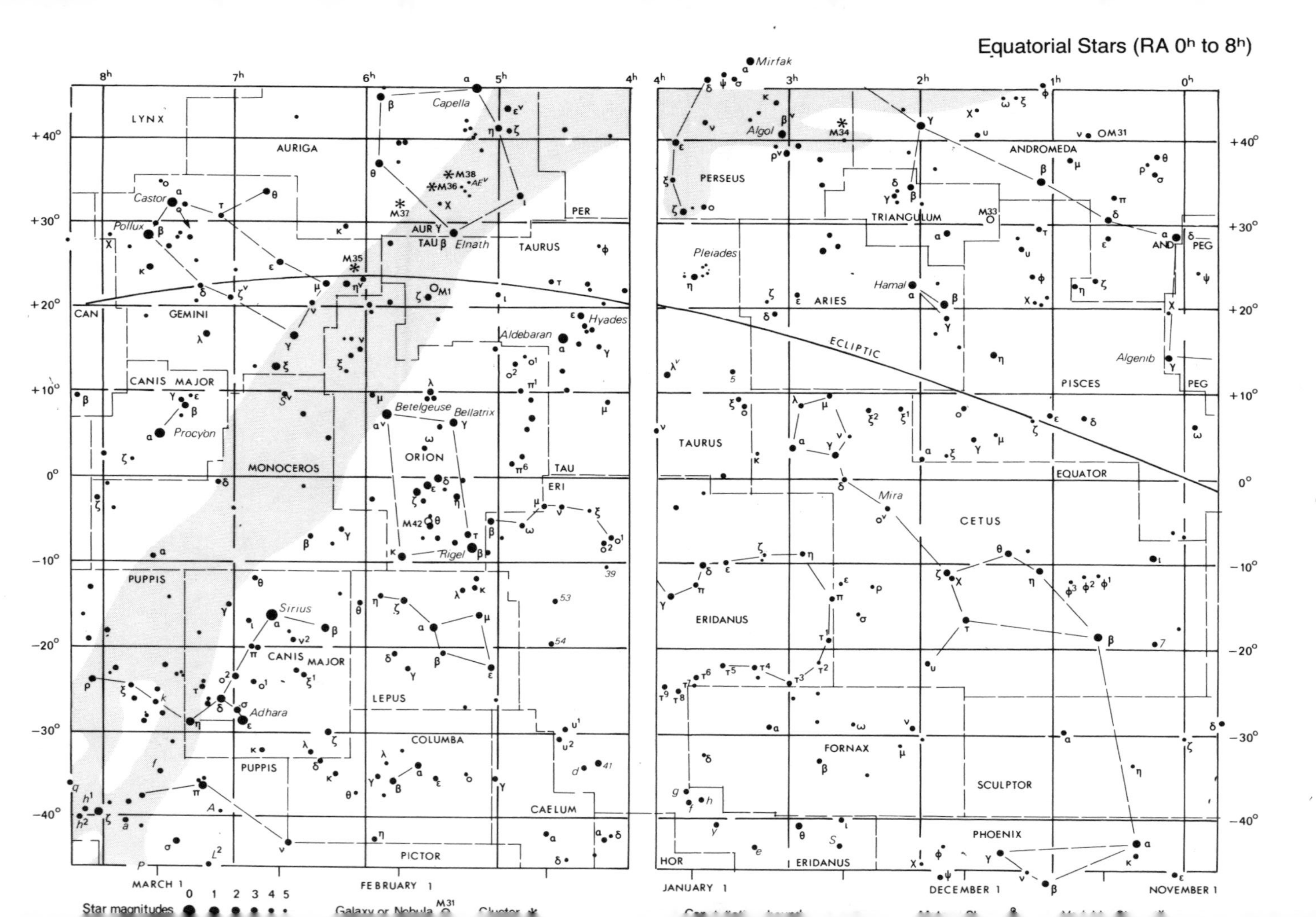
Equatorial Stars (RA 0h to 8h)
LYNX
AURIGA
GEMINI
CAN
CANIS MAJOR
MONOCEROS
ORION
TAURUS
PER
TAU
ERI
PUPPIS
LEPUS
COLUMBA
CAELUM
PICTOR
PERSEUS
ARIES
TRIANGULUM
ANDROMEDA
AND
PEG
PISCES
CETUS
ERIDANUS
FORNAX
SCULPTOR
PHOENIX
HOR
ECLIPTIC
EQUATOR
Capella
Elnath
Castor
Pollux
Procyon
Betelgeuse
Bellatrix
Aldebaran
Hyades
Rigel
Sirius
Adhara
Mirfak
Algol
Pleiades
Hamal
Algenib
Mira
MARCH 1
FEBRUARY 1
JANUARY 1
DECEMBER 1
NOVEMBER 1
Star magnitudes 0 1 2 3 4 5

Star Map 3

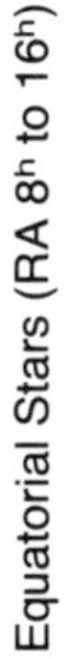
Equatorial Stars (RA 8h to 16h)

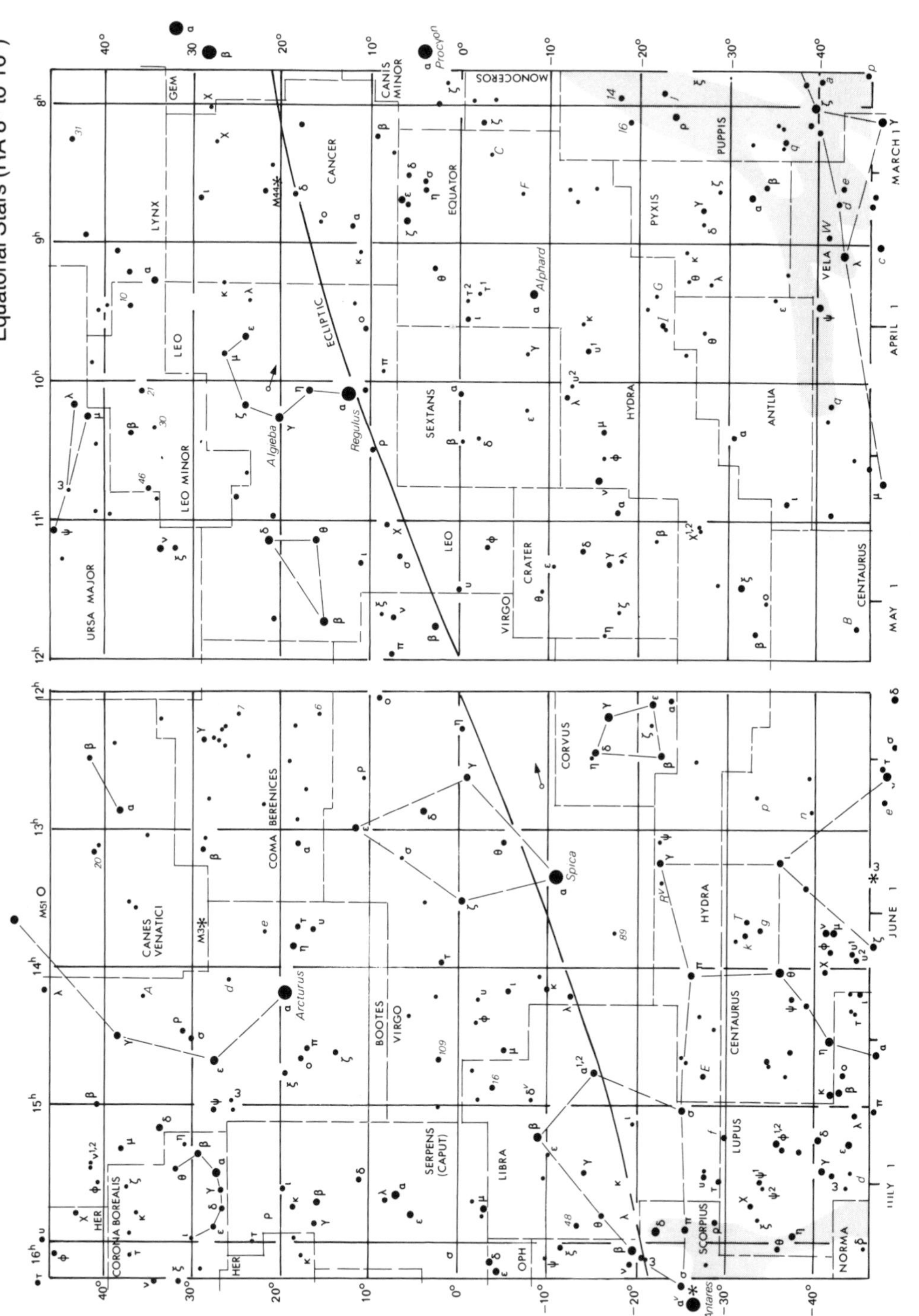
LYNX
LEO
LEO MINOR
URSA MAJOR
CANCER
CANIS MINOR
MONOCEROS
SEXTANS
HYDRA
CRATER
VIRGO
PUPPIS
PYXIS
ANTLIA
VELA
CENTAURUS
GEM
EQUATOR
ECLIPTIC
Procyon
Regulus
Algieba
Alphard
M44
MARCH 1
APRIL 1
MAY 1
CANES VENATICI
COMA BERENICES
BOOTES
CORVUS
Spica
Arcturus
M3
M51
SERPENS (CAPUT)
LIBRA
CORONA BOREALIS
HER
OPH
SCORPIUS
Antares
LUPUS
NORMA
JUNE 1
JULY 1

Star Map 4

Equatorial Stars (RA 16^h to 0^h)

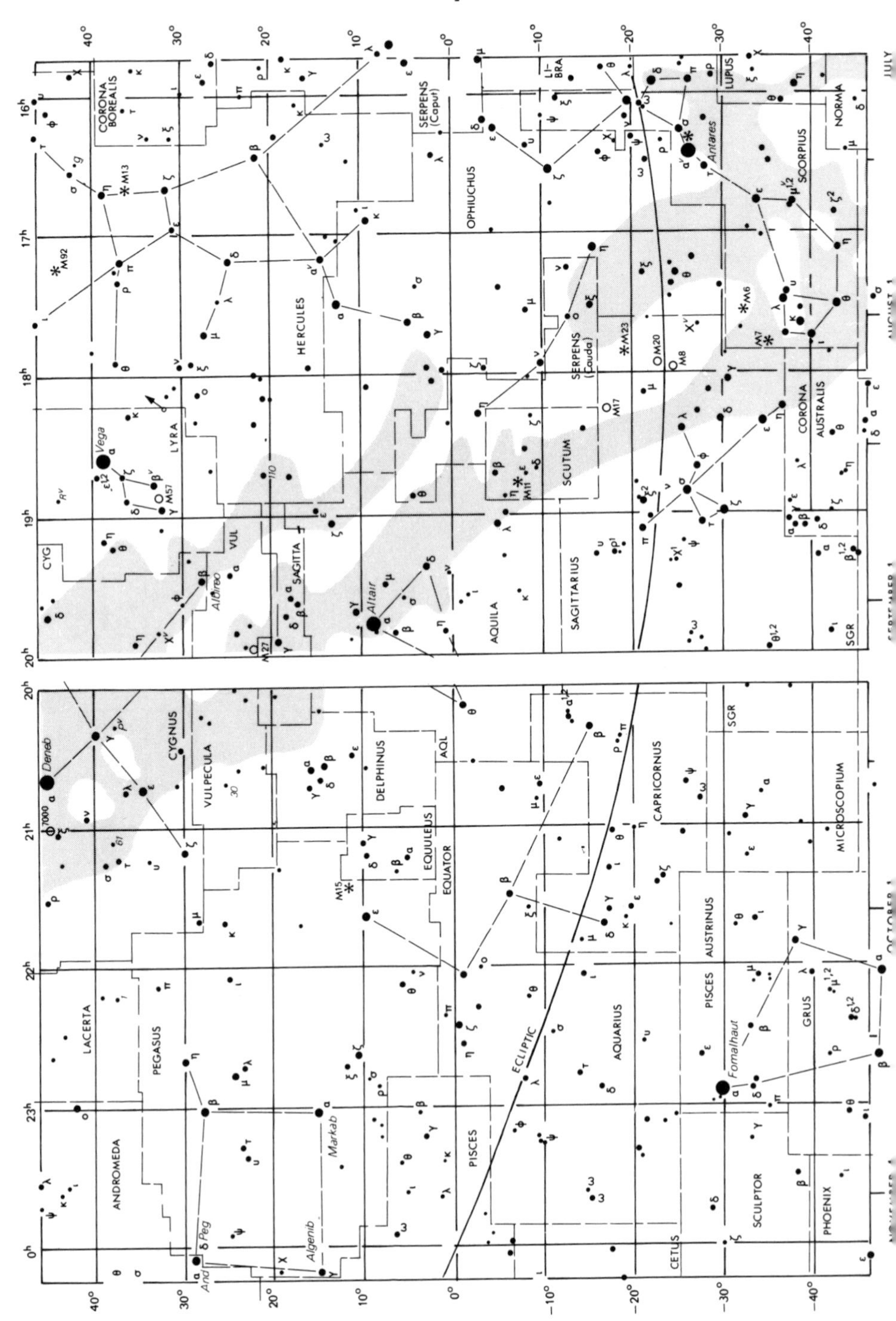
CORONA BOREALIS
SERPENS (Caput)
LI-BRA
LUPUS
NORMA
SCORPIUS
Antares
OPHIUCHUS
HERCULES
M13
M92
LYRA
Vega
M57
CYG
VUL
SAGITTA
Albireo
M27
Altair
AQUILA
SERPENS (Cauda)
SCUTUM
M11
M16
M17
M20
M8
M23
M6
M7
SAGITTARIUS
CORONA AUSTRALIS
SGR
CYGNUS
Deneb
VULPECULA
DELPHINUS
AQL
EQUULEUS
EQUATOR
M15
CAPRICORNUS
MICROSCOPIUM
ECLIPTIC
AQUARIUS
PISCES AUSTRINUS
Fomalhaut
GRUS
LACERTA
PEGASUS
ANDROMEDA
Markab
Algenib
PISCES
CETUS
SCULPTOR
PHOENIX
And
δ Peg
JULY
AUGUST 1
SEPTEMBER 1
OCTOBER 1
NOVEMBER 1

Star Map 5

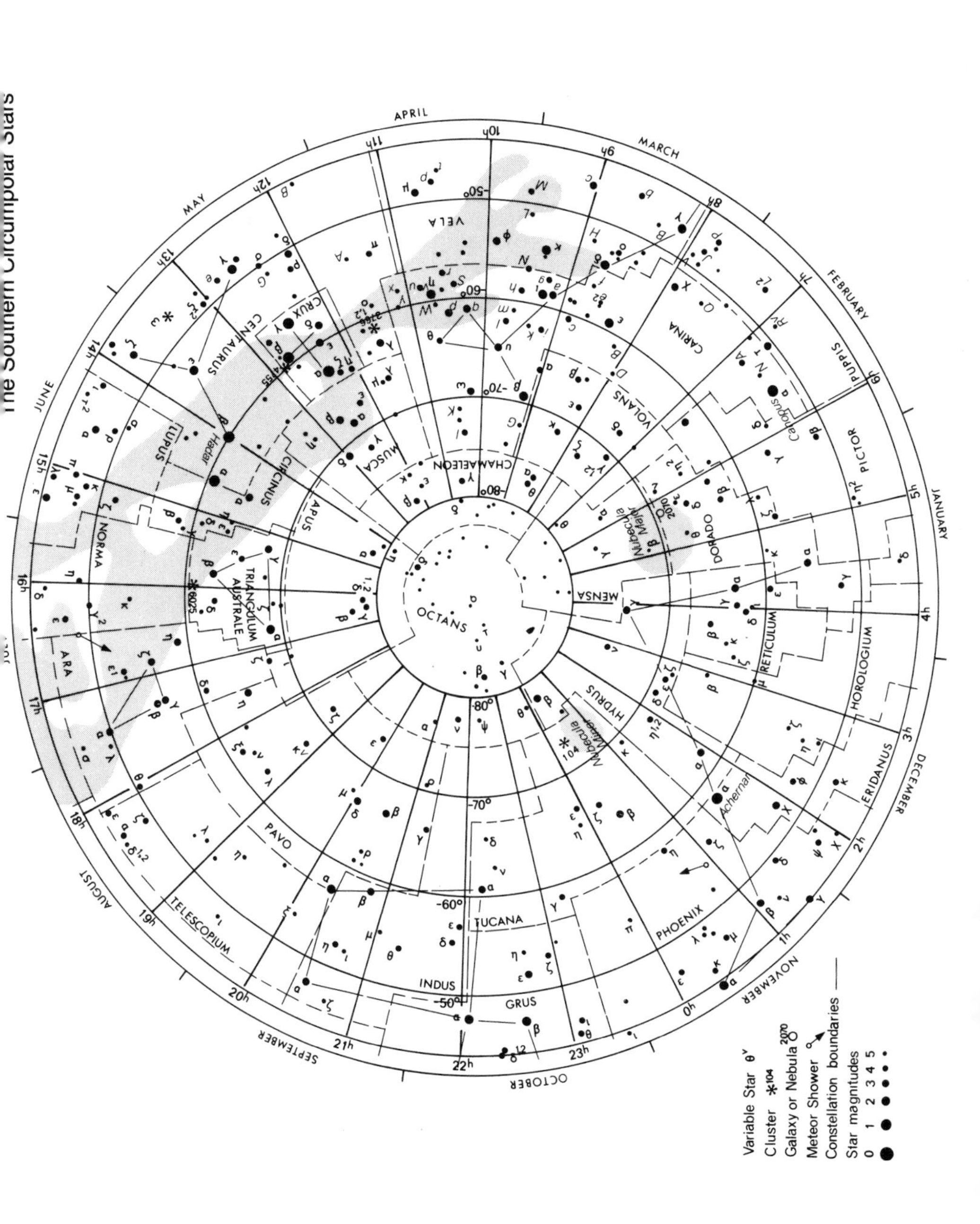
Variable Star
Cluster
Galaxy or Nebula
Meteor Shower
Constellation boundaries
Star magnitudes
0 1 2 3 4 5
APRIL
MARCH
FEBRUARY
JANUARY
DECEMBER
NOVEMBER
OCTOBER
SEPTEMBER
AUGUST
JUNE
MAY
VELA
CARINA
PUPPIS
PICTOR
Canopus
VOLANS
DORADO
Nubecula Major
MENSA
RETICULUM
HOROLOGIUM
ERIDANUS
Achernar
HYDRUS
Nubecula Minor
PHOENIX
TUCANA
GRUS
INDUS
PAVO
TELESCOPIUM
ARA
NORMA
TRIANGULUM AUSTRALE
OCTANS
APUS
CIRCINUS
LUPUS
Hadar
CENTAURUS
CRUX
MUSCA
CHAMAELEON

Exploring with Binoculars and Telescopes

Eyes, Binoculars, and Telescopes

The human eye, binoculars, and telescopes are *all* optical instruments. They have one thing in common: they have lenses or mirrors that *collect* light which they then bend, or focus.

The larger the lens or mirror, the more light-gathering power it has.

The naked eye collects enough light to let us see stars as faint as magnitude 6. With a 1-inch telescope we can see stars to mag 9; a 2-inch telescope reaches stars of mag 10.5; a 3-inch to mag 11.5 (see Star Brightness, or Magnitude).

The 200-inch telescope at Mount Palomar Observatory in California has a visual limit of about mag 19 to 20, but equipped with special electronic devices it can photograph objects as faint as mag 25.

The lens of the human eye is formed by the pupil (the round central part) of the eye, but its size is controlled by the iris (the colored area surrounding the pupil). In bright daylight the pupil contracts down to about two-sixteenths of an inch, but in darkness it expands, or dilates, to five-sixteenths or slightly more. For this reason it is necessary to allow time for our eyes to adjust to the dark when we first go outdoors at night.

The more our pupils expand, the more light they collect. However, human eyes, even when fully dilated, are poor light collectors compared with an ordinary pair of 8x30 prismatic binoculars. Such binoculars have an 8-times magnifying power and a front lens measuring 30 millimetres (1³⁄₁₆ inches). This gives them a collecting power *forty times* greater than the pupils of the human eye.

Even so, the human eye is a truly wonderful creation and a superb instrument in its own right. With our eyes we have a much wider view of the heavens than any binoculars or telescopes can give us. For enjoying, in a single sweep, the panorama of the night sky they are unsurpassed by anything yet invented!

We are not certain *who* invented the telescope. It is usually credited to a Dutch spectacles-maker, Hans Lippershey, in the year 1608, but it is possible telescopes were known in England some time before this. However, they were first applied to looking at the stars when, in 1609, Galileo heard about the new Dutch spyglasses and built one for himself.

Galileo's telescope had a simple arrangement of lenses like those found in opera glasses. The picture, or image, it gave was the right way up as in ordinary field glasses, prismatic binoculars, and terrestrial telescopes used for bird watching.

It was Johannes Kepler, a Bohemian astronomer, who suggested a modification to Galileo's design. In Kepler's telescope, the picture, or image, was upside down. Kepler realized that, in astronomy, it was no disadvantage to see things upside down. His lens design was more efficient and is still used today in astronomical telescopes. This is why many astronomical telescopes show objects upside down.

However, it is still possible to use them for ordinary viewing by employing a special erecting eyepiece that turns the image the right way up. Erecting eyepieces absorb more light and are therefore not as good at revealing faint objects.

All the earliest telescopes used a simple glass lens at the front end. These simple glass lenses were a nuisance, for they produced rainbow colors around an object. In 1668, to overcome this color defect, Isaac Newton designed a new kind of telescope, borrowing and improving upon an idea first thought of by a Scotsman, James Gregory. Seven years earlier Gregory had supposed one could make a telescope using a curved mirror to collect the light instead of using a front lens.

Newton's design was simpler than Gregory's, and the idea proved successful and color-free. In 1672, N. Cassegrain, a Frenchman, designed a similar kind of telescope. This reflected the light back through a hole in the mirror as Gregory's design had done.

Galileo and the first two telescopes he made in 1609.

Isaac Newton and the reflecting telescope he invented around 1668. Also illustrated is the famous apple which, when Newton saw it fall from a tree, supposedly inspired him with ideas about gravity.

Today Newton's, Gregory's, and Cassegrain's ideas are still used in many reflecting telescopes, including the 200-inch at Mount Palomar.

The problem with glass lenses was finally solved in 1733, and now good-quality lenses do not show color fringes. *Achromatic* lenses, as they are called, are used in all modern prismatic binoculars and telescopes. However, the older type of opera and field glasses found in secondhand stores *will* show color fringes (known as chromatic aberration). If no others are available, do not despair. They are good enough to begin with.

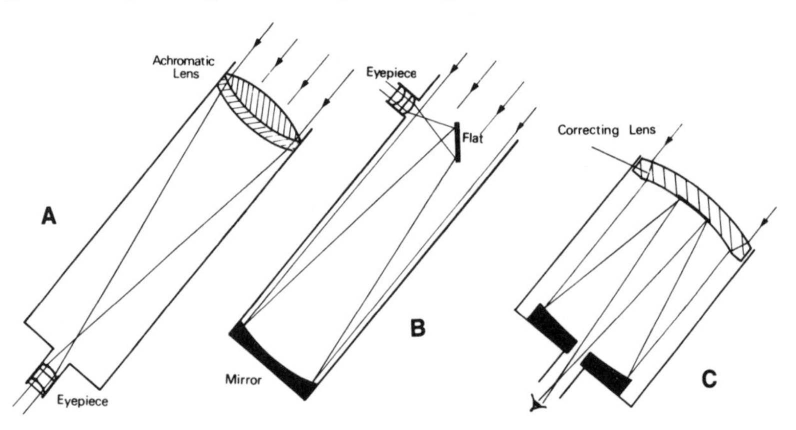

Three popular types of astronomical telescopes:

A An astronomical refractor. See also photo on page 47.
B An astronomical reflector. See also photo on page 47.
C An astronomical telescope that combines refractor and reflector principles. See also color plate 1.

Choosing Binoculars

Usually opera glasses, field glasses, and prismatic binoculars are hand held, but larger binoculars require a small tripod to steady them. A magnification of about x8 (eight times normal size) is the highest practical limit for hand-held instruments. This is because any tremor in your hands is also magnified eight times.

Another factor to consider is the field of view. The higher the magnification, the *narrower* is the view of a starfield. Wider views are preferable as objects are then much easier to locate.

For ordinary binoculars you can buy an inexpensive binocular clip which allows you to attach them to a camera tripod. This is a useful device as it both steadies the binoculars and leaves your hands free to handle a book or a star map.

Older-type opera and field glasses usually have low magnifications, from about x2 (twice normal size) to x6 (six times normal size). They are made on the same simple optical principles as Galileo's first telescopes. Modern prismatic binoculars have a little more magnification and larger fields of view (you see more sky). Binoculars will usually have (blue-tinted) coated lenses which let through more light. These are preferable to the older binoculars that were made before coated lenses were invented.

Typical of modern, wide-angled (wide-view) binoculars is the 7x35 model. These I count among my own favorite binoculars. They give a field of view of 11°30′ which, in terms of the night sky, means that I can see all four stars that form the bowl of the Big Dipper (Ursa Major).

Typical binoculars and field glasses useful for sky-watching.

Any prismatic binoculars in the range 6x30, 8x30, 7x35, 8x40, 7x50, 8x50, or 10x50 will prove satisfactory if the optics are reasonable. Even miniature sporting binoculars like 7x24 and 8x24 can outperform the older-style opera and field glasses, and skywatchers have discovered comets using them.

Binoculars from the Far East are usually much cheaper than domestic or European models, and their quality is reflected by how much you pay. However, some inexpensive Russian binoculars, like their 8x30 model, are of very high quality. I have used such a pair for twenty-five years and I can recommend them.

You can often find "half" binoculars, called monoculars, in sporting and surplus stores, and these are cheaper. My favorite monocular has a 2-inch front lens and a magnification of x7 and performs like a small, very compact, telescope. With this instrument I tracked Halley's famous comet for several months when it revisited the Sun in 1985–86.

Astronomical Telescopes

Astronomical telescopes usually cost more than small binoculars. Ordinary terrestrial or naval spy-glass telescopes *can* also be used for skywatching. However, unless you can fit one with an astronomical eyepiece, its normal erecting eyepiece will absorb a lot of light so that views of fainter starfields and galaxies will be less satisfactory. Some zoom telescopes also share this problem.

But even with terrestrial or naval telescopes you will be able to see craters and mountains on the Moon. They will also give you a glimpse of Jupiter's four brightest satellites and Saturn's rings.

If you plan on making a new purchase, beware of *very* cheap astronomical telescopes. These can easily be recognized. They come with flimsy, rickety mountings, and their cartons advertise high powers of magnification.

Such telescopes are cheap imports and are often sold at supermarkets, general discount stores, and in non-specialist mail-order catalogues—sometimes as special offers. They come illustrated with stunning pictures of the Moon, planets, and galaxies. They give the promise, without actually saying so, that you can actually see such features in *their* telescopes.

If you buy one, you will be disappointed. If your finances are

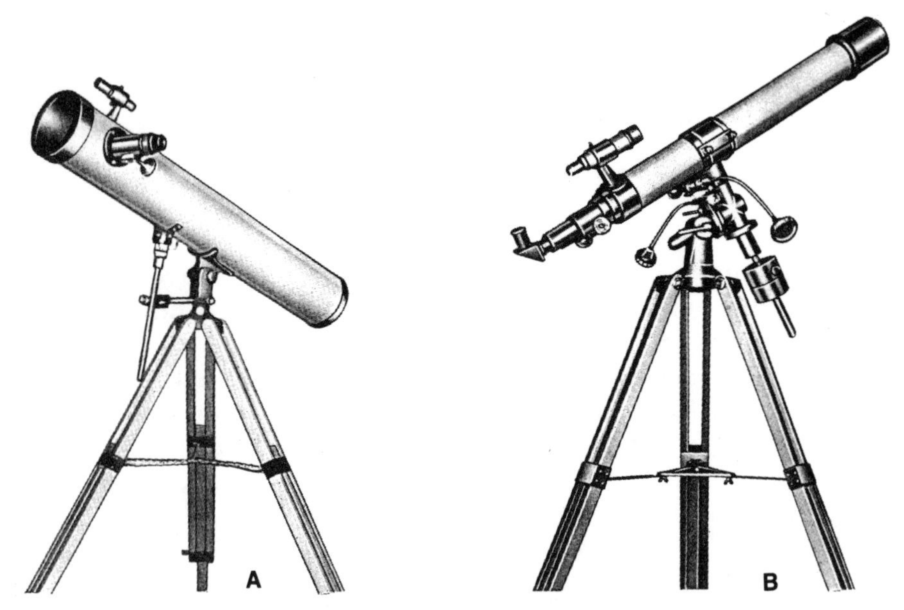

A small astronomical reflector (**A**) and a refractor (**B**) from a supplier's brochure.

A The reflector has a 4-inch-diameter mirror and is equipped with three eyepieces giving magnifications of ×42, ×88, and ×168. It also has a small finder telescope to assist in the quick location of an object.

B The refractor has a (front) object glass 3 inches in diameter and is equipped with four eyepieces giving magnifications of ×36, ×72, ×101, and ×157. Note the star diagonal attached to the eyepiece. A diagonal attachment allows a view of the stars overhead without having to crane your neck.

limited, buy some inexpensive binoculars instead and save up for a proper astronomical telescope.

When you can afford a proper astronomical telescope, either new or secondhand, visit a specialist optical store or, alternately, your local pawnbroker. There are genuine secondhand bargains to be had if you tread cautiously and perhaps take someone more knowledgeable along to help negotiate.

To find such a store, look in the Yellow Pages of the phone book. If there is no store locally, try the mail-order firms which advertise in popular astronomical magazines. In these magazines you have a wide choice and you can send for their brochures to study at leisure before you buy.

Top-class astronomical telescopes do not come cheap, but you do not have to buy the very best as a trainee skywatcher.

The most suitable kinds of telescopes for beginners are *refractors* rather than *reflectors*. For beginners, reflectors can be tricky to use and maintain. Refractors are easier to lift and carry about.

The most portable telescopes, but rather expensive, are the Questar or Celestron types. They are simple to operate and maintain and are ideal for those skywatchers who have no garden or backyard to observe from. These types are widely advertised in popular astronomical magazines.

For a beginner with less money to spend, the ideal telescope is a refractor with a front lens (called an objective, or object glass) measuring 2 to 3 inches in diameter. It should be equipped with a star diagonal (to make viewing overhead less of a strain on your neck) and, if possible, *three* interchangeable eyepieces providing magnifications of approximately x25, x50, and x75 or x100.

If you can only afford a telescope with a single eyepiece, make sure it is one of low magnifying power, say x25 to x35. Higher magnifying powers, although necessary for lunar and planetary detail and for splitting close double stars, restrict the field of view. Higher powers are for nights when the air is steady. When you are magnifying an object, you are *also* magnifying the turbulence in the atmosphere, and on breezy and frosty nights the image will be dancing about.

If you can afford it, equip yourself with both binoculars *and* a telescope. Seasoned skywatchers are like golfers and have a range of favorite "clubs." For some celestial objects one "club" performs better than another.

A cluster of distant galaxies in Corona Borealis. Galaxies, like stars, seem to favor family groups. Around the sky 1,000,000 galaxies have been counted down to mag 19. Still fainter galaxies exist in numbers uncountable.

Stars and Galaxies

The Natural History of Stars

When looking at the brighter stars, you will see that they differ in color.

If the following stars are visible, notice that Betelgeuse, in Orion, is a bright orange color while Rigel, a close neighbor, is bluish-white. Compare the yellow-orange color of Arcturus, in Bootes, with the pure whitish color of Vega, in Lyra, and the white color of Altair, in Aquila. Compare the dull red of Antares, in Scorpio, with the blue of Spica, in Virgo.

You can make these observations with the naked eye, but if you have binoculars, try them out on the above stars. Through binoculars, these colors are more obvious.

The color differences between stars are real ones. They vary because stars differ in their chemical makeup, size, and age.

We know that the color of a star is a direct indication of its temperature. Blue stars are hot, but red stars are cooler bodies. Sol, our own star, is a yellow color with a temperature between hot stars and the cooler ones.

Like us, all stars are born. They grow up, mature, age, and then die. Of course, the time scales for humans and for stars are vastly different. The interval between the birth and death of a star is several billion years.

Stars differ in size, behavior, and temperature. Some stars are enormous giants with diameters over 3,000 times larger than the Sun's. Other stars are dwarfs which emit so little light we can barely see them with the largest telescopes. There are even some stars which emit no visible light, and we have to detect them by other means.

Some stars are cool globes of tenuous gas that slowly pulsate like enormous hearts, expanding and then contracting in a regular rhythm over several days or months. These are called variable stars, and we can watch their pulsations by their changes in brightness, or magnitude.

Other stars are small, very dense planetlike bodies where the force of gravity is so great that an object the size of a pea would weigh nearly a ton. A future astronaut visiting such a star would be immediately squashed flat.

All stars shine by the burning of their nuclear furnaces. They operate like gigantic power stations, pumping out light and heat over their very long life spans. Some stars, during their lives, go out of control. Their rate of burning increases and they suddenly blow up, breaking apart into myriad fragments. These are called exploding stars, or novae (see below).

It is perhaps reassuring for us to know that Sol, our Sun, is a user-friendly nuclear power station and a very average stable kind of star. It is likely to keep on shining and burning steadily for several more thousand million years.

Star Brightness, or Magnitude

Even the most casual glance at the night sky reveals that the stars differ in brightness.

Some stars like Sirius, in Canis Major, flash like brilliant diamond points. Others are so dim we can barely see them.

The brightness of a star is called its apparent magnitude, often written mag for short.

Stars differ in brightness because of two things: first, their true brightness, or luminosity; second, how far away they are. If two stars are of equal luminosity, but one is farther away, the one farther away will appear less bright to us and its apparent magnitude will be fainter.

It was the ancient Greek astronomer Hipparchus who first decided that the stars might conveniently be divided into six grades of brightness. The very brightest he called 1st magnitude stars; the next brightest 2nd magnitude stars, and so on to the 6th magnitude. A 6th magnitude star is the faintest star we can see

The Brightest Stars

Star	Apparent Magnitude	Distance (Light-Years)	Color
Sirius (α Canis Majoris)	–1.47	8.7	Brilliant-white
Canopus (α Carinae)	–0.71	181.0	Yellow-white
α Centauri	–0.1	4.3	Yellow
Arcturus (α Bootis)	–0.2	35.86	Yellow-orange
Vega (α Lyrae)	0.1	26.4	Blue-white
Capella (α Aurigae)	0.2	45.64	Yellow
Rigel (β Orionis)	0.3	880.0	Blue-white
Procyon (α Canis Minoris)	0.5	11.4	Yellow-white
Achernar (α Eridani)	0.6	114.1	Blue-white
β Centauri	0.9	423.8	Blue-white
Betelgeuse (α Orionis)	0.7	586.0	Reddish-orange
Altair (α Aquilae)	0.9	16.4	White
Aldebaran (α Tauri)	1.1	68.46	Yellow-orange
β Crucis	1.5	260.8	Blue-white
Antares (α Scorpii)	1.2	423.8	Reddish
Spica (α Virginis)	1.2	211.9	Blue-white
Pollux (β Geminorum)	1.2	34.8	Yellow-orange
Fomalhaut (α Piscis Austrini)	1.3	22.82	White
Deneb (α Cygni)	1.3	1,630.0	White
α Crucis	1.1	423.8	Blue-white

with the naked eye on clear, moonless nights in dark country areas.

This simple system worked quite well until the invention of the telescope, but thereafter it soon proved unsatisfactory.

Eventually, in the nineteenth century, the system was put on a more scientific footing. It was decided that the difference between one magnitude and the next would be a step of 2.5 times. This means that a star of magnitude 1 is 100 times brighter than a star of magnitude 6.

Under the modern system very bright stars, like Sirius, which used to be called a 1st magnitude star under the old system, needed to be reclassified. The very brightest stars like Sirius and Canopus now have minus magnitudes (Sirius is mag –1.47 and Canopus mag –0.71). Note that magnitudes are often given to a hundredth part of a mag, but in normal skywatching one-tenth of a mag is more the custom. With practice, the human eye can detect differences of one tenth of a magnitude fairly easily. Most star lists, like the ones here, provide magnitudes to the nearest tenth.

The brightness of other celestial objects is also given in star magnitudes. On the same scale, the planet Venus often shines at mag –3.0, so at its brightest it outshines Sirius. The Moon, when full, is of mag –12, and the Sun at midday is of mag –27.

The Magic in Sun and Starlight

Isaac Newton's greatest discovery was finding how gravity worked in keeping the planets in orbit around the Sun and the Moon around the Earth. However, another of Newton's discoveries came as a result of his experimenting with a beam of sunlight.

Newton found, to his surprise, that when he passed a beam of sunlight through a wedge-shaped prism of glass, it was split up, or refracted, into various colors—the very same colors we see when we look at a rainbow.

Newton did not carry his own experiment much further than this, but later astronomers, repeating what Newton had done, found something else. They discovered that some of the rainbow colors—red, orange, yellow, green, blue, indigo, and violet—were crossed by mysterious dark lines which later were called absorption lines, or bands.

For a time, these lines, or bands, were a great puzzle. Then two German scientists, G. R. Kirchhoff and R. W. Bunsen—the inventor of the Bunsen burner—solved the mystery. They found that these lines were produced by substances we knew on Earth, such as iron, copper, hydrogen, and sodium. The lines were, in fact, chemical printouts of substances that also formed the Sun.

Since the Sun was a star, astronomers now looked at the stars in the night sky and discovered the same lines. By passing starlight through a prism, it was possible to find out what stars were made of. It was *almost* magical!

Studies of these chemical lines, or bands, that occur in the spectrum of the Sun and other stars is called *spectroscopy*. The modern device containing the prism (or these days more often a ruled grating which performs like a prism) is called a *spectroscope*. It is attached to a telescope at the focal end.

After the invention of the telescope, the spectroscope was the next important invention to help in the discovery of the true nature of stars. A beam of starlight literally carries with it the life history of the star. Through the spectroscope we can decipher how hot a star is, how big it is, how old it is, and where its future lies.

Variable Stars

Not all stars shine with a constant brightness. Some, like Algol, "the Demon Star," in Perseus, vary over an interval of a few hours. Others, like the redgiant star Betelgeuse, in Orion, vary over a year or more. Some, like Algol, have variations as regular as clockwork while Betelgeuse slowly wanes and waxes to no fixed times.

The large group of stars known as variables are divided into about five main types. Some of them are not true variables. These are the spectroscopic binaries like Algol which only vary their light because there is more than one star in the system. One star periodically passes in front of the other, eclipsing its companion (see also double stars).

Variable stars provide astronomers with important clues about how stars evolve over their long lifetimes. It is a branch of astronomy where amateur skywatchers can play a useful role, adding to the work of the professionals. In addition to carrying out regular magnitude estimates of known variable stars, there are amateur skywatchers who are constantly on the lookout for novae, or temporary stars, often called "new" stars. These are not really new stars as such, but previously faint stars which suddenly "explode," increasing their brightness by as much as 50,000 times.

Several telescopic novae occur every year. Some novae are bright enough to be seen with the naked eye about once every two or three years. Fame can be the reward for the amateur skywatcher who first detects a nova and reports it to an authority. No one can tell when the next nova will explode in *your* sky. But to be able to spot one you must get to know all the stars intimately. As chance would have it, as I write this, in February 1992, I have a report in front of me of a mag 4.2 nova in Cygnus which has just been spotted by an amateur in Colorado.

Do not expect to find a "new" star on your first night outdoors. You may, of course, have beginner's luck as others have had in the past. In the long run, however, it is dedication that brings results.

A "new star," or nova, discovered in Delphinus on July 8, 1967, by an amateur skywatcher. Note the diamond-shaped star pattern below the arrowed nova. This star pattern can be identified on Star Map 4.

Double, or Binary, Stars

After the invention of the telescope, astronomers discovered that some stars which appeared single to the naked eye were really two or more stars close together.

Some of these stars appear close only because they happen to be lying in the same line of sight as we view them from Earth. In reality, they are separated from each other by many light years.

Other stars, however, were found to be true pairs, even triples or more, that revolve in orbits around a common center.

The true double stars are called binary stars. Binary is a Latin word meaning “two.” We now know of systems where five or six, or more, stars are involved. These are still called binary systems because the name these days refers to stars which have a physical connection.

Thousands of binary systems have been recorded. Among the stars closest to the Sun, at least half turn out to be binaries. It is as if some stars, like humans, prefer to live in close family groups. This rule probably holds good throughout the Milky Way galaxy.

It has been suggested that our own Sun has a companion star, one with an orbit which periodically carries it far away. This has never been proved, but some astronomers keep up their search for it.

Many binary systems are visible through small telescopes. A few of the brighter, wider, pairs are within range of binoculars. One famous double system, ε (Epsilon) Lyrae, lies close to the bright star Vega. If your vision is good, it can just be glimpsed with the naked eye, but it is plainly visible with the smallest optical means.

Some binary systems, however, are so close that even the largest telescopes will not show them except as a single star.

How then do we know they *are* binary systems?

There are two ingenious ways of detecting them.

Because *both* binary stars are moving in their respective orbits around a common center of gravity, this movement shows up as a tiny “wobble” that can be measured on photographs taken over intervals of time. If a “wobble” is detected, it indicates that the “single” star is a double system. Many binaries have been discovered this way.

The second method involves the spectroscope (see above). Although the light from two stars very close together appears in a telescope as a *single* point, its spectrum will show the chemical lines *doubled* owing to the presence of two stars. In this respect, the spectroscope is a better detective than the telescope.

The spectroscope is an even better detective in other ways, too. By carefully measuring the shifts in position of the two sets of chemical lines as the stars swing around in their orbits, it is possible to work out the size of their orbits as well as how massive and hot each star is. All double stars so detected are called spectroscopic binaries.

Star Clusters

All the stars we see with the naked eye, binoculars, and small telescopes belong to the Milky Way, our own galaxy.

Some of these stars are gathered together in groups called clusters. There are two kinds: globular star clusters and open star clusters.

The globular clusters are the more impressive objects. Most of them lie at a great distance from the Sun and are vast collections of stars numbering some hundreds of thousands. The stars are so

> Following the invention of the telescope, generations of skywatchers searched the heavens to discover new objects beyond reach of the naked eye. They made lists, or catalogues, of those they found interesting.
>
> Many such catalogues exist, but perhaps the most famous is that by the Frenchman Charles Messier.
>
> Messier was interested in finding new comets. During his sweeps of the heavens, using a very small telescope, he came across scores of fuzzy, cloudlike objects which at first glance could be mistaken for comets. He therefore decided to make a list, or catalogue, of all these nuisances.
>
> Today we recognize these "nuisances" as star clusters, nebulae, and distant galaxies; and Messier's list of 103 objects represents the brighter deep-sky objects visible in binoculars and small telescopes.
>
> Messier's numbers are still in use today to identify these objects, and it is customary to refer to them simply by their M number. For example, the Great Globular Cluster in Hercules was number 13 on Messier's list, so we refer to it as M13. The Andromeda galaxy is M31.

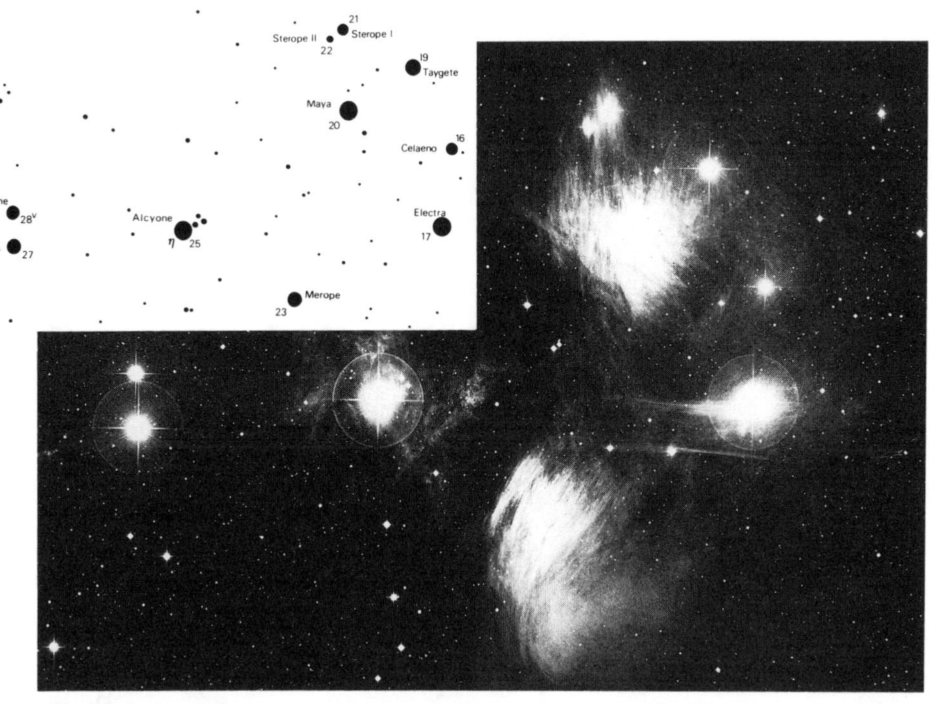

The Pleiades star cluster, or the Seven Sisters. With the naked eye most people can see only six stars, but even the smallest binoculars reveal many more (see inset star map and Star Map 2).

numerous that, when they are photographed with large telescopes, they seem to merge at the center into a dense ball of light.

One spectacular globular cluster is M13 in Hercules (see color plate 20), visible in binoculars or small telescopes as a fuzzy star. Another example, suitable for viewing in the southern hemisphere, is ω (Omega) Centauri. This cluster is even brighter than the one in Hercules and is visible to the naked eye, shining like a dull star of magnitude 3.7.

Open clusters, in contrast to globular clusters, are less-densely packed groups of stars with little or no crowding together at their centers. They may consist of a few hundred stars, but often less, and they are generally arranged quite haphazardly. Perhaps the best known among the open clusters are the Pleiades, a group called "the Seven Sisters." This beautiful cluster of stars is plainly visible to the naked eye in the constellation of Taurus. Some people with good eyesight will be able to see all Seven Sisters, but most of us can only see six stars with the naked eye. Even with small binoculars, scores of other stars become visible. The Pleiades open cluster is one of the showpieces of the sky.

Another bright open cluster is Praesepe (M44), in Cancer, known as the Beehive star cluster. This too can be seen without optical means, appearing to the naked eye as a hazy patch of light. It provides an excellent starter object for beginners.

Spotting the Milky Way

Outdoors, when identifying the constellations, you may also notice, if the street lights do not cause too much glare, a faint band of light stretching across the heavens. Its position in the sky will depend on the time of year (see the Star Maps).

This band of light extends over more than one-tenth of the visible night sky and represents the Milky Way, the galaxy of stars to which the Sun and its family members, including ourselves, all belong. The Milky Way surrounds us on all sides.

Most of the distant stars forming the Milky Way lie in a flatish disc-like plane, and we see them edge on. They are so far away we cannot see them as separate points of light, and they merge to suggest a "milky" appearance, hence the name the Milky Way. An ancient Greek poem refers to it as "that shining wheel, men call it Milk." Since milk in Greek was *gala,* it was named the Galaxy.

It is calculated that 400 billion stars belong to the Milky Way

The Milky Way extends across 25 separate constellations. If we start from Scorpius, it traverses Sagittarius, Scutum, Aquila, Sagitta, Vulpecula, Cygnus, Lacerta, and projects an arm into Cepheus. It then passes through Perseus, Auriga, between Gemini and Orion, then Monoceros, and Canis Major.

In the southern skies it extends across Puppis, Pyxis, Vela, Carina, Musca, Centaurus, Lupus, Circinus, Norma, Ara, and then back into Scorpius.

In built-up areas polluted by street lights and smoke, the Milky Way can rarely be seen at its best as the ancients once saw it. If you ever get a chance to visit a desert region, look out for the Milky Way. You will feel you can almost reach out and touch it.

galaxy. From end to end it spans 100,000 light-years. If you were a space traveller looking back at the Milky Way from outside, it would look very much like we see the Andromeda galaxy, known as M31 (see color plate 22).

We know that the entire Milky Way is slowly turning on its axis, which is located somewhere beyond the star clouds in Sagittarius. The speed of the rotating Milky Way varies according to how far out an object lies from its center. The Sun and its family members lie two-thirds of the way out from the center, and we take 200 million years to go around once.

The Galactic Nebulae

In addition to discovering star clusters, the early telescopic sky-watchers came across other curious bodies that could not be resolved into separate stars but remained hazy and cloudlike. For this reason such a body was called a nebula, a Latin word meaning "cloudlike" or "misty."

The North America nebula (NGC 7000). This gaseous nebula lies near Deneb in Cygnus (see Star Map 4). On a very clear, dark night it can be seen in binoculars as a faint glow.

When telescopes became larger, it was found that some nebulae were very distant globular clusters, but other nebulae remained cloudlike.

Although the name nebula is still in common usage today, it describes several different kinds of cosmic bodies. The true nebulae, the galactic nebulae, lying *inside* the Milky Way, are known to be glowing clouds, or rings, of gas and dust. The term galactic simply means "belonging to the Milky Way."

When photographed with large telescopes, some of these glowing gas clouds, or rings, appear to have distinct shapes which remind us of familiar things. For example, the North America nebula does remind us of the shape of North America, and the Veil nebula looks like a veil (see color plate 13). Another famous example is the Horsehead nebula in Orion (see color plate 15).

The constellation of Orion is host to one of the brightest of the gas and dust nebulae, M46. It can be glimpsed with the naked eye as a hazy spot adjacent to θ (Theta) Orionis. Even in small binoculars it will reveal its distinct greenish color due to the presence of oxygen in the gas. The Great Nebula in Orion, as it is often called, is believed to be a cosmic nursery where new stars are created from the condensing gas and dust clouds.

Some other galactic nebulae are called planetary nebulae, so called because they have the superficial appearance of round planetary bodies. They are shells of gas blown off from a dim central star. One or two bright planetary nebulae can just be glimpsed with small binoculars as faint starlike bodies. Most require a 2- to 3-inch telescope to show their form. The most celebrated planetary nebula is the Ring nebula (M57) in Lyra (see color plate 12). In photographs it looks exactly like a smoke ring.

Other strange nebulae are dark objects—so dark they look like holes in the Milky Way. The most famous one lies in the southern hemisphere and appears so dark it has earned the nickname "the Coal Sack." These nebulae are not really holes in the starfield, but vast clouds of obscuring dust which lie in the plane of the Milky Way and blot out the stars behind.

The Magellanic Clouds

In the southern hemisphere, there are two very interesting naked-eye patches of distant stars similar to but separate from the Milky Way. These are the Magellanic Clouds.

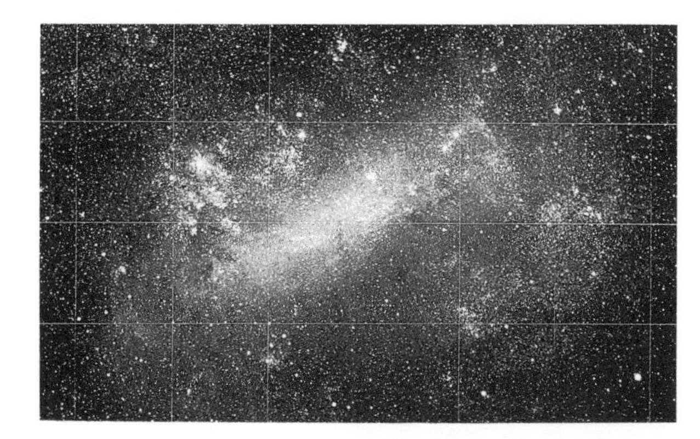

The Large Magellanic Cloud, Nubecula Major (see Star Map 5).

The Small Magellanic Cloud, Nubecula Minor (see Star Map 5).

They were named after the Portuguese navigator Ferdinand Magellan, who, in 1519, set out to sail around the world. He was one of the first to take notice of these curious cloudlike bodies. However, long before Magellan they were known to the Arab astronomers, who called the Larger Cloud *Al Bakr,* the White Ox.

As signposts in the sky they are just as important as the Southern Cross. In Australia, they were once known as "the Drovers' Friends." They were used as a sure directional aid by outback cattlemen when stock were herded and moved in the cool of night.

The Larger, or Greater, Cloud lies sprawled across the constellations of Dorado and Mensa. The Smaller, or Lesser, Cloud is located in nearby Tucana. On the star maps, you will see them labelled by their scientific names: Nubecula Major (the Larger Cloud) and Nubecula Minor (the Smaller Cloud). The name nubecula comes from the Latin *nubes*, meaning "little clouds."

Although rather smaller than our own Milky Way, both clouds

are independent galaxies in their own right. They are closely related to us in space as members of what are known as the local group of galaxies.

The Larger Cloud lies 150,000 light-years away, and the Smaller Cloud 165,000 light-years away. It is possible that once in the remote past they both formed part of our own Milky Way system but became detached.

The Larger Cloud contains the star S Doradus. It is probably the most luminous star known in our region of the universe.

In 1987, the Larger Cloud suddenly became the focus of world attention. On the night of February 23 of that year, a brilliant "new star," of a type known as a supernova, burst into view.

It was first spotted independently by several keen-eyed astronomers including the veteran amateur skywatcher Albert Jones, a New Zealander, who has watched the southern stars over a lifetime.

Back in 1946, Jones discovered a new comet that now bears his name. Over the years he came to know the southern skies so well he could spot a stranger in an instant.

Supernovae are not rare events in distant galaxies, but they are very rare in the Milky Way. The supernova sighted in 1987 was the first one to occur nearby for nearly 400 years, since one appeared in the constellation of Ophiuchus in 1604. This latter star is often called Kepler's Star, but it was actually first seen by his keen-eyed pupil John Brunowski. History is not always correct in assigning discoveries.

The supernova found in the Larger Cloud is now designated SN 1987A—meaning a supernova discovered in 1987. The A signifies it was the first one seen in that year.

This supernova has been studied by astronomers more than any other star in the heavens. Investigations have shown it to be a massive blue star as well as a pulsar. Pulsars are very special kinds of stars that were not known until the middle 1960s.

Stars like SN 1987A are thought to be the "factory" stars which manufacture those chemicals in the universe that make up parts of the Earth and other planets and provide the source material for life. Some pulsars are believed to have planetarylike bodies revolving around them. Several such discoveries have been reported in recent years. At least one pulsar has several other bodies going around it which might be similar to planets.

Beyond the Milky Way

Following Charles Messier's list of 103 nebulae in 1781, the numbers soon increased when bigger telescopes came into use. Many of the new nebulae were discovered by the German-born William Herschel, a young bandsman who quit his home country and travelled to England to start a new career.

He became the greatest amateur skywatcher of all time.

In England, Herschel totally immersed himself in astronomy. Being unable to afford a telescope, he decided to make one for himself. His experiments were with the reflecting-type telescope of the kind Isaac Newton had invented. He soon found he could make better telescopes than any the British had made before him. He made scores of them during the next few years and became world famous for his skill.

He became even more famous for his observations of the heavens.

In 1781 he discovered the planet Uranus during one of his sky sweeps, but it was really the mysterious nebulae that fascinated him. By 1802 he had observed and then catalogued 2,500 of them.

In his observations he was assisted by his sister Caroline, who had followed him to England. She became a very skilled skywatcher in her own right and discovered six comets.

The Herschels built bigger and better telescopes than anyone before them. His largest had a mirror 48 inches in diameter, but most of his work was done using smaller instruments. He was able to penetrate farther into space than anyone before him, and he began to suspect that some of the nebulae were distant, independent systems of stars which lay beyond the Milky Way. Perhaps they were "island universes," but neither Herschel nor anyone else could be sure.

After Herschel's death, his son, John, carried on his work,

This British anniversary stamp shows Sir William Herschel and his giant 4-foot reflecting telescope. Herschel is the figure on the left holding a chart showing the orbits of the moons of Uranus.

travelling to South Africa to observe the southern skies. There he discovered nebulae his father had been unable to see from the north.

In 1845 the Earl of Rosse built a large telescope at Birr, in Ireland. It had a mirror 72 inches in diameter. When this was turned on the nebulae, some of them showed details not visible in Herschel's telescopes. The object in Messier's catalogue numbered as M51, which was seen by the Herschels to consist of rings of nebulous matter, now showed an intriguing "whirlpool" form (see color plate 14). Modern photographs taken in large telescopes show this feature remarkably well.

In the Depths of Space

No one could be sure that the M51 Whirlpool nebula and the many others like it, which showed spiral forms, were distant galaxies similar to our own Milky Way. They were so far away that astronomers had no method by which they could measure their remoteness.

Some astronomers thought the nebulae were objects just inside the boundaries of the Milky Way. Others thought them to be much farther away. Were they perhaps at distances of some tens of thousands of light-years beyond the Milky Way? No one could tell.

In 1918 a new telescope was turned on the nebulae in the hope that it would solve the riddle. This telescope was the 100-inch telescope at Mount Wilson, in California. In the hands of two largely self-taught skywatchers, Edwin Powell Hubble and Milton Humason, the nebulae slowly began to give up their secrets.

Another American astronomer, Harlow Shapley, had realized that a kind of pulsating variable star, known as a Cepheid (pronounced seefid), might be used in a very ingenious way to measure how far away the spiral nebulae lay.

Using the 100-inch at Mount Wilson, Hubble succeeded in photographing some faint Cepheids in various spiral nebulae. Adapting the ideas of Shapley, he came up with a figure that astounded some astronomers. If his assumptions were correct, some of the spiral nebulae were at least *one million light-years away.*

This was a truly astonishing discovery. It proved that the Herschels and others, who had called some of the nebulae "island

1 A Questar, high-performance telescope. This very compact and portable telescope can be used visually or with a camera, as shown.

2 A safe method of projecting the image of the Sun to view sunspots.

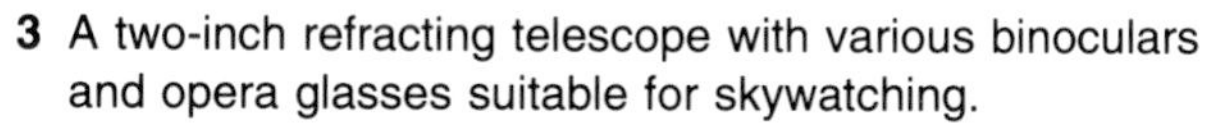

3 A two-inch refracting telescope with various binoculars and opera glasses suitable for skywatching.

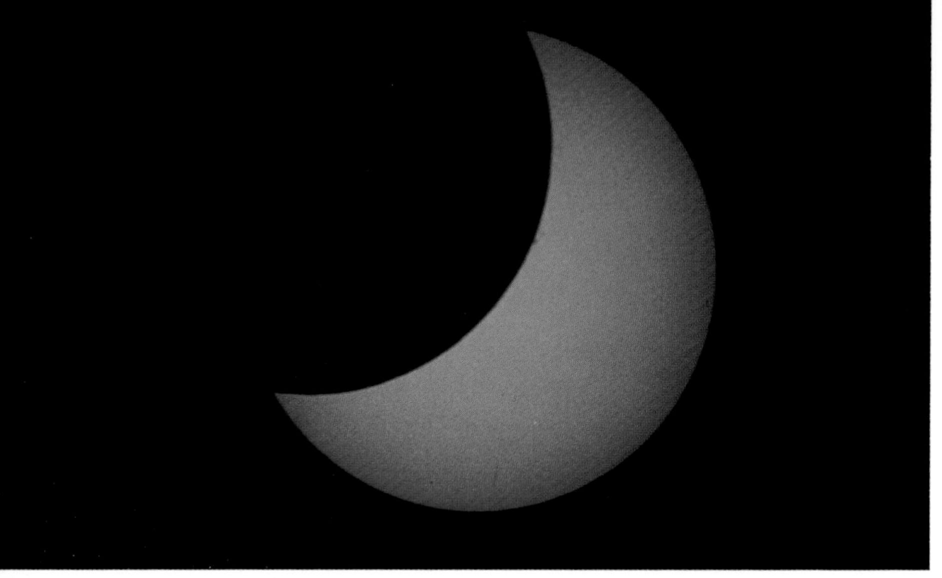

4 A partial eclipse of the Sun.

5 A total eclipse of the Sun.

6 The diamond-ring effect is caused by the first rays of the reappearing Sun after a total solar eclipse.

7 Saturn and its rings. The planet's cloud belts, just visible, are much less prominent than those on Jupiter.

8 The Sun with some small surface spots.

9 An eclipse of the Moon showing the curve of the Earth's shadow.

10 Jupiter and its cloud belts showing the Great Red Spot.

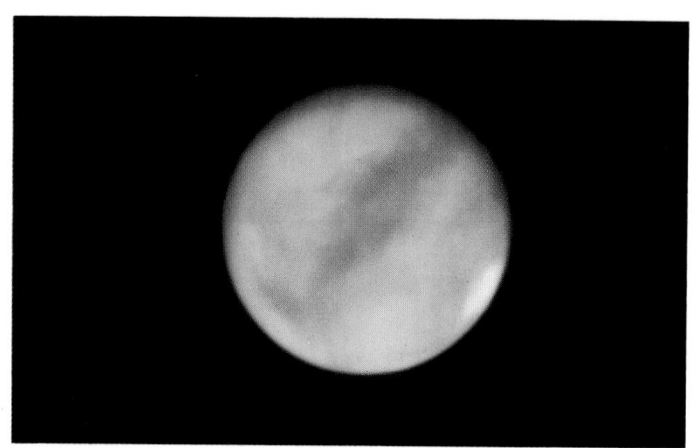

11 Mars showing dark surface features and the south polar cap. These can be seen in small telescopes when Mars is close to the Earth.

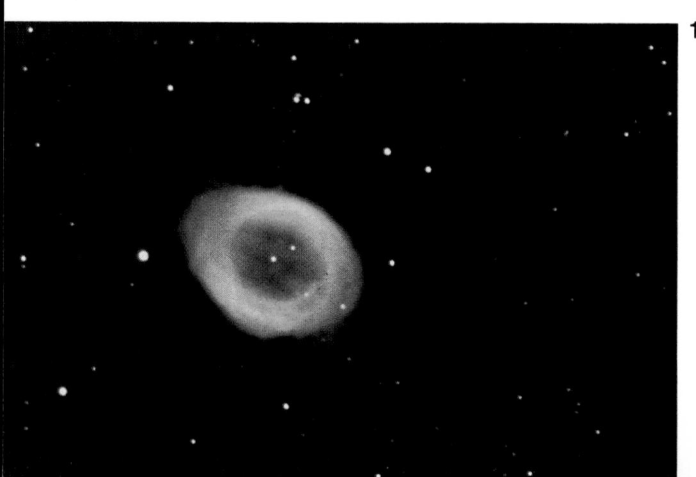

12 The Ring nebula (M57) in Lyra. In small telescopes, this magnitude 9.3 planetary nebula appears like a small hazy star (see Star Map 4).

13 The Veil nebula (NGC 6992) in Cygnus was probably created by a supernova explosion some 50,000 years ago.

14 The Whirlpool galaxy (M51) in Canes Venatici. At magnitude 8.1, it appears in binoculars as a faint blob of light (see Star Map 1).

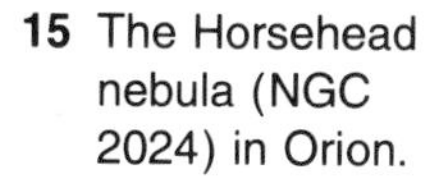

15 The Horsehead nebula (NGC 2024) in Orion.

16 The Aurora Borealis, or Northern Lights, in action.

17 A track of an artificial satellite above the Earth crosses the great star cloud in Sagittarius, which lies in the center of the Milky Way galaxy.

18 Halley's comet seen from Australia in March 1986. Below the Milky Way, Halley's 10° tail points towards four bright stars, Zeta, Tau, Sigma, and Phi Sagittarii (see Star Map 4).

19 Donati's spectacular comet, as seen over Paris in 1857. The bright star near the head of the comet is Arcturus (see Star Map 3).

20 The globular cluster (M13) in Hercules. In binoculars it appears as a faint, misty patch of light to the right of the "flower pot" shape of Hercules (see Star Map 4).

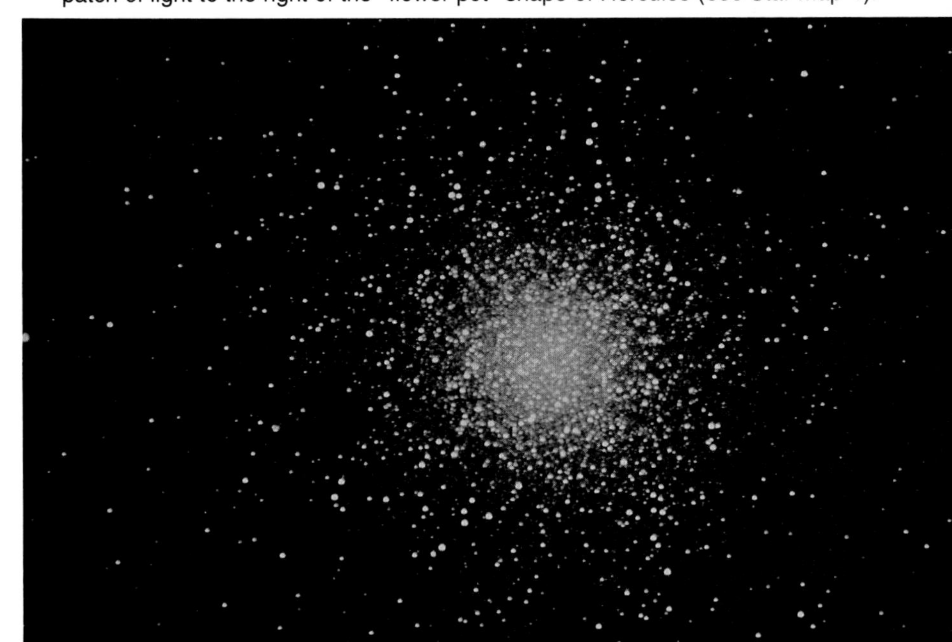

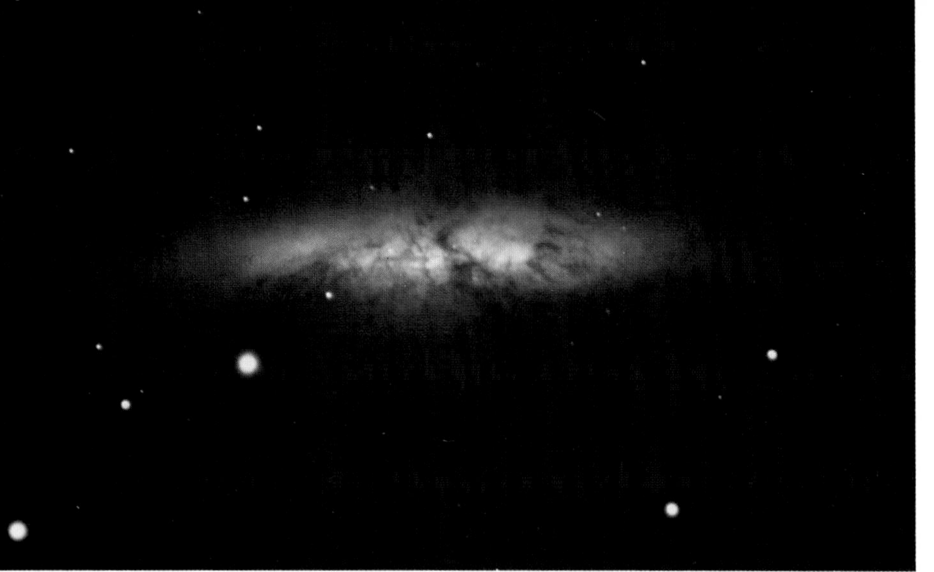

21 An irregular galaxy (M82) in Ursa Major. M82 and its companion spiral galaxy (M81) are both just within range of binoculars and are seen in the same field of view (see Star Map 1).

22 The Great Nebula (M31) in Andromeda, the nearest of the larger galaxies. It is plainly visible to the naked eye as a misty patch of light (see Star Map 2).

23 A spiral galaxy (NGC 7331) in Pegasus.

universes," had been right. But more surprises were in store, for in later years Hubble's distances needed to be corrected. They had to be *doubled*!

Further observations of these distant galaxies, as they were now known to be, revealed some more remarkable things. Their spectrums showed them to be rushing away, or receding, from us at tremendous speeds. It gave birth to the idea that the whole universe is expanding. The farther one penetrates into space, the faster the galaxies are retreating. The more distant ones are rushing away from us at the speed of many thousands of miles every second.

Large modern telescopes now find galaxies in such numbers they outnumber the background stars of the Milky Way. Like stars, galaxies swarm in clusters and the clusters themselves form superclusters.

Quasars—at the Edge of the Universe?

Astronomy is an exciting science, for new discoveries are being made all the time. One quite unexpected discovery was the quasars—a name coined in the early 1960s to describe a new class of object.

Years before, some faint, bluish stars had been photographed, but no one paid them much attention. However, when radio telescopes came into service, they began to pick up some powerful radio signals, visually identified as coming from star-like objects. Some of these turned out to be the bluish stars known about before.

Because these stars emitted powerful radio signals, whereas other stars emitted little or no signals, astronomers were puzzled by them. In 1963, Maarten Schmidt, a Dutch astronomer using the 200-inch telescope at Mount Palomar, decided to photograph the spectrum of one of the blue radio stars. He chose the brightest one, a 12.8 magnitude star in Virgo known as 3C-273. (Like stars and nebulae, radio-source objects are listed in various catalogues. The designation 3C-273 means that this object is number 273 in the 3rd Cambridge Catalogue of Radio Sources.)

Schmidt, following in Hubble's footsteps, hoped to find some clues among the chemical lines of the star. He did, and was astounded. If his interpretation of 3C-273's spectrum was correct, he realized he had found a velocity redshift far greater than anything seen before. This faint, bluish "star" was rushing away from us at 15 percent the speed of light!

Other bluish stars (now renamed quasars, coined from quasi-stellar, meaning *almost* starlike) were examined, and even bigger redshifts discovered. One of them, 3C-9, was found to be receding at 80 percent of the speed of light and was thought to be 10 billion light-years away. Since Albert Einstein, in his theory of relativity, had found that no object could exceed the speed of light, astronomers wondered if they were now peering towards the edge of the visible universe.

Today, no one can be sure what the quasars really are. They might be peculiar starlike bodies or small galaxies—or something in between.

What is certain is that quasars emit tremendous amounts of energy, and most of them show large redshifts. A large redshift in a spectrum has always indicated that a body was moving away from us at high speed. But could the redshifts mean something else? If redshifts indicate high speeds away from us, there are probably quasars so far away and travelling so fast that their light never catches up with us!

Some astronomers now query the very nature of the redshifts found in quasars. They wonder if they may be due to something else—perhaps the effects of very strong gravity or that light itself behaves differently when making a long journey through space. Some have gone so far as to deny that quasars are very distant objects but are instead located at the edge of our own Milky Way.

In the future, bigger telescopes may provide the final answer.

Finding Quasar 3C-273

Unlike ordinary galaxies, which may be seen with small telescopes and some even with hand-held binoculars, quasars are very faint eyes-on objects. The brightest one, 3C-273, in Virgo, shines like a star of magnitude 12.8. If you want to glimpse what may be the brightest of the more distant objects in the universe, you will need a good 8-inch telescope.

Even if you do not have such a telescope, you can still view that *region* of the sky. The quasar is located about 3½° north-east of the star η (Eta) Virginis and 1½° west of the galaxy NGC 4536, which is detectable with a 4-inch telescope.

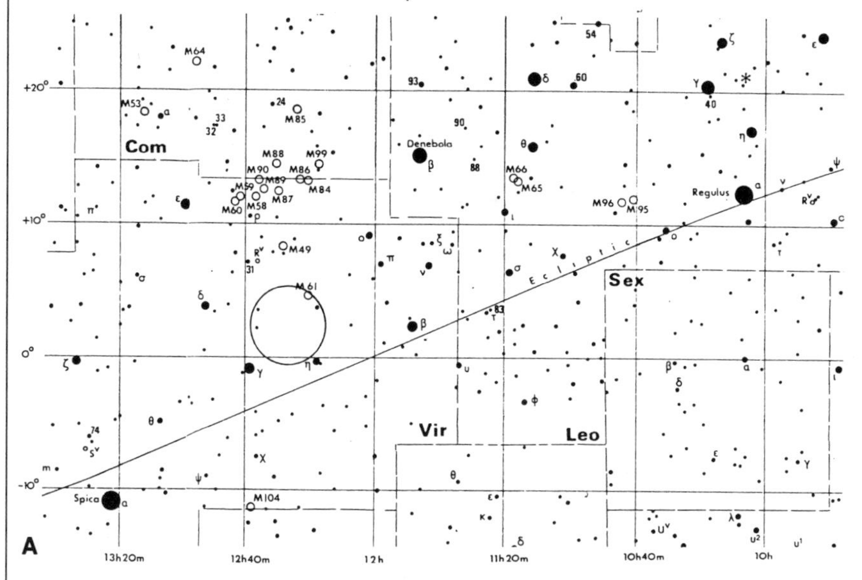

Finder star maps for Quasar 3C-273.

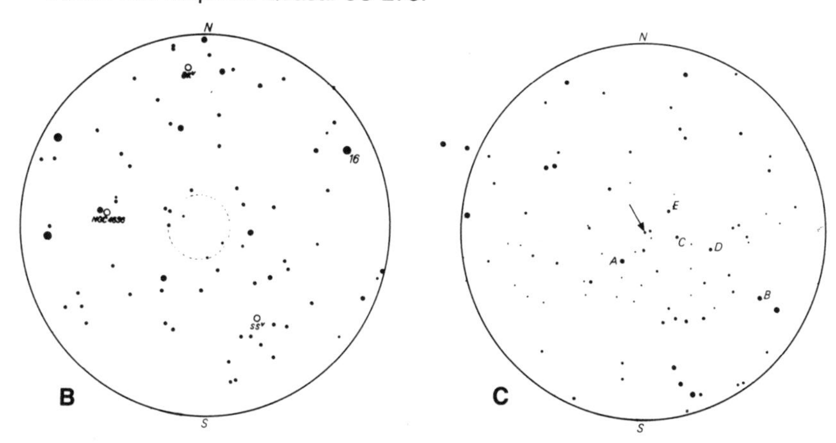

A Naked-eye stars nearest the Quasar (circled).

B The surrounding stars as seen in binoculars and small telescopes. The field of view is approximately 5½°.

C Quasar 3C-273 arrowed. The field of view is approximately 1°. Comparison star mags are as follows: star A = 10.1; star B = 10.4; star C = 11.9; star D = 12.3; star E = 12.6; NGC 4536 mag = 11.8. The faintest stars shown are about mag 15.

The spectroscope, which provided the first direct evidence about the chemistry of the stars, today plays a key role in our understanding of the universe. In the nineteenth century it was found that when a celestial object was moving towards us, its chemical lines were shifted towards the blue end of its spectrum and when it was going away from us (receding), the lines were shifted towards the red end of its spectrum.

In the 1920s, when the spectrums of many galaxies were photographed, nearly all were found to have redshifts.

Astronomers have concluded that the redshift of the galaxies indicates the universe is expanding. It is expanding in all directions. The farther we travel outward, the greater the redshift, and the faster the universe is expanding. The largest redshifts yet discovered are found in the very peculiar quasars, and most astronomers now believe these are the most distant bodies known to us.

Black and White Holes

Black holes refer to regions in space where the pull of gravity is believed to be so strong that light is prevented from escaping. It is a region where matter from the outside is drawn into and swallowed up by a cosmic "whirlpool." These black holes are not to be confused with the Coal-Sack type of black holes we see in the Milky Way due to obscuring dust clouds blotting out the stars behind.

Even with the largest telescopes we cannot "see" a black hole directly—as eyes-on objects they do not exist! But astronomers have good reasons for believing that they do. They can be detected, indirectly, by the effects of their gravity and the emission of X-rays nearby.

Some believe that black holes are common features, and that there is probably one at the center of the Milky Way and in the centers of other galaxies.

White holes are not as well known as black holes. These are also peculiar objects thought to exist. However, instead of matter being drawn in, as in black holes, matter is thrown out of a white hole. Perhaps every black hole has a complementing white hole linked to it.

Neither black nor white holes are yet *proven* observational facts. They are imaginative ideas to explain how light and matter behave in distant space. Confirmation awaits new discoveries.

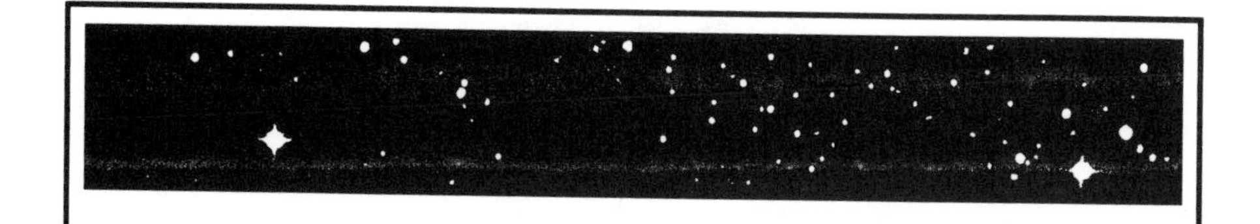

Deep Skywatch

Hints on Using Binoculars and Telescopes

What stars, star clusters, nebulae, and galaxies you see with a pair of binoculars or a telescope depends on the size of the instrument's front lens or mirror (the aperture). The larger the aperture, the fainter the celestial object you will be able to see.

A rough guide to the faintness seen in different apertures is given below:

Size of front lens or mirror	Naked eye	1 in.	2 in.	3 in.	4 in.	6 in.	8 in.	12 in.
Closest double star split min mag x50 (approx) (″ = seconds of arc)	180″	4½″	2¼″	1½″	1¼″	¾″	½″	⅓″
Faintest (star) magnitude seen	6	9.0	10.5	11.4	12.0	12.9	13.5	14.4

In splitting double stars, ordinary binoculars perform less well than telescopes because a higher magnification is required. Low-power binoculars, magnifying in the range x6 to x12, will not usually split double stars less than 150″ apart.

Finding Faint Objects

To observe faint star clusters, nebulae, galaxies, and comets you need very clear, moonless nights, free from sky haze and the glare of street lights.

If you live in a town or city, position yourself in the dark shadow of a building. This helps to reduce the intrusive glare from nearby street lights. To take in the whole sky you will have to be prepared to move about. Choose for preference a lightweight portable instrument that is easy to carry and set up again.

When searching for star clusters, nebulae, galaxies, and comets, start with the lowest magnifying power. This will give you a larger (wider) field of view, and the object sought will be easier to spot.

First let your eyes become fully dark-adapted. From the star maps choose and then *identify* a nearby bright star to use as a signpost.

Estimate how many fields of view the faint object is away from the bright star you have identified. This can be measured off in degrees using the vertical declination scale on the star maps.

Starting from the bright star, *slowly* sweep towards the faint object, moving back and forth.

You will soon notice that it is easier to spot a faint object out of the corner or side of your eyes than straight on. This is because the sides of your eyes are more sensitive to light than the middle parts.

Finding the Field of View

You can check out the fields of view of your binoculars and the different eyepieces of your telescope by looking at the Full Moon.

The Full Moon's apparent diameter is 30′(½°). Reckon how many times you can fit the Moon into your view, and this will give you the field of view in degrees and minutes.

Another method of finding the field of view for binoculars is by looking at some key stars.The distance between Dubhe and Merak, the Pointer stars in Ursa Major, is 5°. In Orion's Belt the distance between Delta and Zeta is 1¼°. The distance between Alpha and Beta Centauri is 5°.

Knowing the approximate field of view helps you in sweeping up a faint object.

Astronomers call this using averted vision. By using averted vision, a surprising number of star clusters, nebulae, and galaxies can be glimpsed through small binoculars—once you know *exactly* where to look.

Naked-eye Variable Stars

Variable		Position RA		Position DEC		Magnitudes MAX	Magnitudes MIN	Period (days)	Type
α	Cas	00h	39m	+56°	16′	2.5	3.1		Irr
γ	Cas	00	54	+60	27	1.6	3.0		Irr
o	Cet†	02	17	–03	12	2.0	10.1	331.48	LP
β	Per (A)	03	05	+40	46	2.2	3.5	2.8673	EA
λ	Tau	03	58	+12	21	3.5	4.0	3.9530	EA
ε	Aur	04	58	+43	45	3.7	4.5	9883	EA
AE	AUR*	05	13	+34	15	5.4	6.1		Irr
α	Ori	05	53	+07	24	0.4	1.3	2070	SR
η	Gem	06	12	+22	31	3.1	3.9	233.4	SR
R	Hya†	13	27	–23	01	3.5	10.9	387	LP
δ	Lib*	14	58	–08	19	4.8	5.9	2.3273	EA
α	Sco	16	26	–26	19	1.2	1.8	1733	SR
g	Her*	16	27	+41	59	4.6	6.0	80	SR
μ′	Sco	16	48	–37	58	3.0	3.3	1.4463	β Lyr
α′	Her	17	12	+14	27	3.0	4.0	100	SR
X	Sgr*	17	44	–27	49	5.0	6.1	7.0122	δ Cep
β	Lyr	18	48	+33	18	3.4	4.3	12.9080	β Lyr
R	Lyr	18	54	+43	53	4.0	5.0	50	SR
χ	Cyg†	19	49	+32	47	2.3	14.3	406.66	LP
P	Cyg	20	16	+37	53	3.0(?)	6.0(?)		N
μ	Cep	21	42	+58	33	3.6	5.1		SR
δ	Cep	22	27	+58	10	3.9	5.0	5.3663	δ Cep
ρ	Cas	23	52	+57	13	4.1	6.2		Irr
R	Cas†	23	56	+51	07	4.8	13.6	430.93	LP

Note: SR. Semi-Regular star; A = Algol; EA. Eclipsing Binary star of Algol-type; δ Cep. Delta Cepheid-type variable; β Lyr. Beta Lyrae-type variable; Irr. Irregular-period star; LP. Long-period star; N. Nova-type star. * Difficult naked-eye object. † Naked-eye only at maximum.

Interesting Double Stars

Star and Constellation		Magnitudes	Distance Apart (sec)	Colors	Position		
					RA		DEC
γ	And	3.0, 5.0	10	yellow, blue	02^h	00^m	+42°.1
ι	Cnc	4.4, 6.5	30	yellow, blue	08	44	+20°.0
α	CVn	3.2, 5.7	20	blue, blue	12	54	+38°.6
α^1, α^2	Cap	4.0, 3.8	376	yellow, yellow	20	15	–12°.7
ι	Cas	4.2, 7.1, 8.1	2,7	yellow, blue, blue	02	25	+67°.2
α	Cen	0.3, 1.7	4	yellow, red	14	37	–60°.6
ζ	CrB	4.1, 5.0	6	white, blue	15	38	+36°.8
α	Cru	1.4, 1.9	5	blue, blue	12	24	–63°.8
β	Cyg	3.0, 5.3	35	yellow, blue	19	29	+27°.8
γ	Del	4.0, 5.0	10	yellow, green	20	44	+16°.0
ν	Dra	4.6, 4.6	62	white, white	17	31	+55°.2
ψ	Dra	4.0, 5.2	31	yellow, purple	17	43	+72°.2
32	Eri	4.0, 6.0	7	yellow, blue	03	52	–03°.1
α	Gem	2.7, 3.7	5	white, white	07	31	+32°.0
α	Her	3.0, 6.1	4	orange, green	17	12	+14°.4
ε	Lyr	4.6, 4.9	208	yellow, blue	18	43	+39°.6
ε^1	Lyr	4.6, 6.3	3	yellow	18	43	+39°.6
ε^2	Lyr	4.9, 5.2	2	blue	18	43	+39°.6

Star and Constellation		Magnitudes	Distance Apart (sec)	Colors	Position		
					RA		DEC
β	Ori	4.0, 10.3, 2.5, 6.3	(Quadruple)	blues	05	36	–02°.6
α	Sco	1.2, 6.5	3	red, white	16	26	–26°.3
β	Tuc	4.5, 4.5	26	blue, white	00	29	–63°.2
ζ	UMa	2.4, 4.0	14	white, white	13	22	+55°.2
α	UMi	2.5, 8.8	19	yellow, blue	01	49	+89°.0
γ	Vir	3.6, 3.7	6	white, yellow	12	39	–01°.2

Some Bright Star Clusters

Constellation	Object	Position			Type	Mag	Remarks
		RA		DEC			
Auriga	M 38	05^h	25^m	+35°.8	Open	7.4†	
Auriga	M 37	05	49	+32°.6	Open	6.2†	
Auriga	M 36	05	32	+34°.1	Open	6.3†	
Cancer	M 44	08	37	+20°.2	Open	3.7*	"Praesepe," or "the Beehive"
Canes Venatici	M 3	13	40	+28°.6	Globular	6.4†	
Cassiopeia	M 103	01	30	+60°.4	Open	7.4†	
Centaurus	NGC 3766	11	34	–61°.3	Open	5.1†	
Centaurus	ω	13	24	–47°.0	Globular	3.7*	½° diameter
Crux	NGC 4755	12	51	–60°.1	Open	5.2†	
Cygnus	M 39	21	30	+48°.2	Open	5.2†	
Gemini	M 35	06	06	+24°.4	Open	5.3†	

Constellation	Object	Position		Type	Mag	Remarks
		RA	DEC			
Hercules	M 13	16 40	+36°.6	Globular	5.7†	"Great Cluster"
Hercules	M 92	17 15	+43°.1	Globular	6.1†	
Lacerta	NGC 7243	22 13	+49°.5	Open	7.4†	Near β Lac
Pegasus	M 15	21 28	+12°.0	Globular	6.0†	
Perseus	NGC 869 and 884	02 18	+56°.9	Open	4.4* 4.7*	"Swordhandle Double Cluster"
Perseus	M 34	02 39	+42°.5	Open	5.5†	
Sagittarius	M 23	17 54	–19°.0	Open	6.9†	
Scorpius	M 4	16 20	–26°.5	Globular	6.4†	Near Antares
Scorpius	M 6	17 37	–32°.2	Open	5.3†	
Scorpius	M 7	17 51	–34°.8	Open	3.2†	
Scutum	M 11	18 48	–06°.8	Open	6.3*	
Taurus	M 45	03 44	+24°.0	Open	1.6*	"the Pleiades"
Triangulum Australe	NGC 6025	15 59	–60°.4	Open	5.8†	
Tucana	NGC 104	00 22	–72°.4	Globular	3.0*	Star 47 Tucanae

* Visible to naked eye; † Visible in binoculars.

Galaxy M104, the Sombrero nebula, located in Virgo. This is a galaxy seen edge on. It can be glimpsed in large binoculars and small telescopes when the sky is clear and dark (see Finder Map A on page 67).

Interesting Galaxies and Nebulae

Constellation	Object	Position			Type	Mag	Remarks
		RA		DEC			
Andromeda	M 31	00^h	40^m	+41°.0	Spiral Gal	4.8*	"Great Nebula"
Canes Venatici	M 51	13	28	+47°.4	Spiral Gal	8.1†	"Whirlpool Nebula"
Cygnus	NGC 7000	20	57	+44°.0	Diffuse Neb	—	"North America"
Dorado	NGC 2070	05	39	−69°.2	Diffuse Neb	—*	Visible to naked eye
Lyra	M 57	18	52	+33°.0	Plane-tary Neb	9.3‡	"Ring Nebula"
Orion	M 42	05	33	−05°.4	Diffuse Neb	4.0*	"Great Nebula"
Sagittarius	M 20	17	59	−23°.0	Diffuse Neb	6.9†	"Trifid Nebula"
Sagittarius	M 8	18	01	−24°.4	Diffuse Neb	6.8†	"Lagoon Nebula"
Sagittarius	M 17	18	18	−16°.2	Diffuse Neb	7.0†	"Omega Nebula"
Taurus	M 1	05	32	+22°.0	Diffuse Neb	8.4‡	"Crab Nebula"
Triangulum	M 33	01	31	+30°.4	Spiral Gal	6.7†	Large and faint
Ursa Major	M 81	09	52	+69°.3	Spiral Gal	7.9†	Connected to M 82
Ursa Major	M 82	09	52	+70°.0	Spiral Gal	8.8‡	Connected to M 81
Ursa Major	M 97	11	12	+55°.3	Plane-tary Neb	11.0**	"Owl Nebula"
Vulpecula	M 27	19	58	+22°.6	Plane-tary Neb	7.6†	"Dumbbell Nebula"

* Visible to naked eye; † Visible in binoculars; ‡ Requires 2-inch telescope; ** Requires 3-inch telescope.

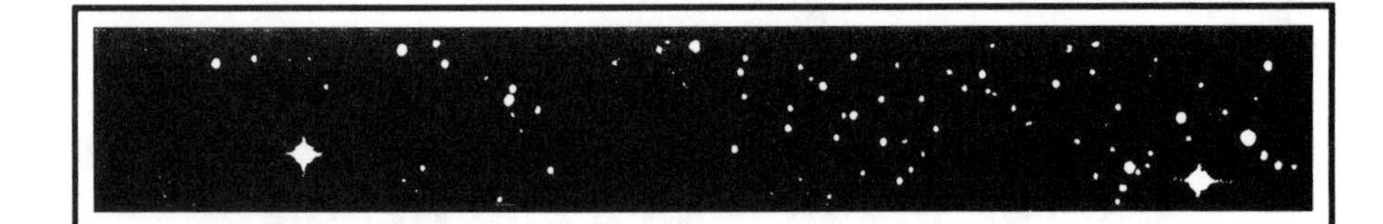

The Solar System

Earlier astronomers noticed that certain bright stars were not fixed but shifted, or wandered, in relation to other stars. These shifting stars were called planets—from the Greek *planetes,* meaning to wander.

The early astronomers counted five bright wanderers. By Roman times they had become known as Mercury, Venus, Mars, Jupiter, and Saturn. These names stemmed from the mythical beings in the ancient world. The other major planets we know of today carry similar mythological names. Uranus was discovered in 1781, Neptune in 1846, and Pluto in 1930.

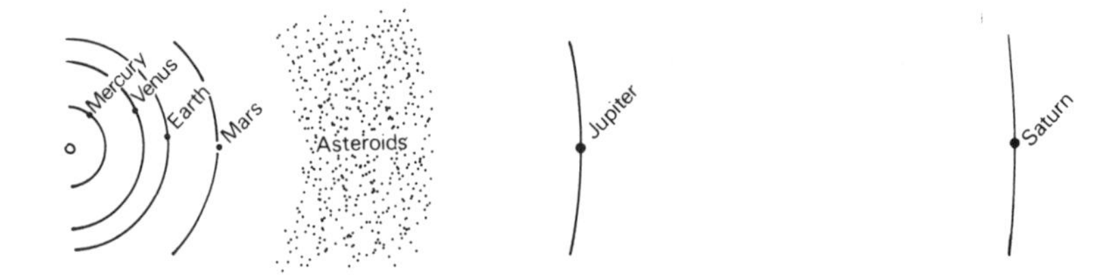

The Sun's family of planets.

In addition to the major planets, there are also thousands of minor planets, or asteroids. The first asteroid, called Ceres, was found in 1801.

Other members of the Sun's family include our own Moon; the moons, or satellites, of all the other planets; comets; shooting stars; and meteorites.

All objects that make up the solar family come under the gravitational pull of the Sun. All revolve around the Sun in different paths called orbits. The orbits of the major planets are almost circular in shape, but the orbits of asteroids and comets are more egg-shaped, or elliptical.

The planets:

A Venus in crescent phase. Because Venus lies between the Earth and the Sun, it shows phases like the Moon does (see drawing on page 90).

B Mars showing faint dark surface markings and one of the polar caps. These features can be seen in small telescopes when Mars is close to the Earth.

C Jupiter and its principal cloud belts. These cloud belts and the famous red spot are visible in small telescopes similar to the one in color plate 1.

D Saturn and its rings. The dark Cassini division is visible in small telescopes when the rings are at their widest openings (see page 97).

E Pluto is so far away and so small in diameter that photographs do not show it as a disk.

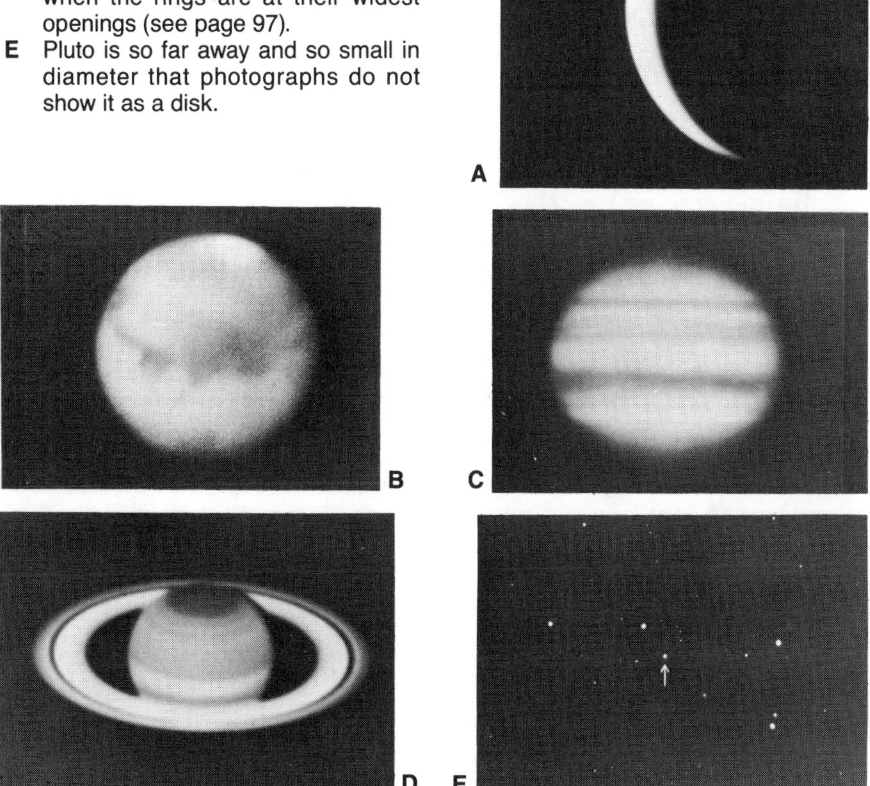

The Sun

Sol, our Sun, is a typical star and is made up of a large globe of gas. The Sun generates all its light and heat from a nuclear power plant deep inside its central core. Because Sol generates its own light, we refer to it as a self-luminous body. The planets, and other satellites, however, do not shine by their own light and are therefore not self-luminous. We can only see them because they reflect sunlight off their surfaces.

Although Sol is only a small star, in comparison with the planets it is a massive body with a diameter of about 865,000 miles. The Sun lies 93 million miles away from us, but even though it is so far away, the Earth soaks up solar energy at the rate of 5 million horsepower for every square mile of our surface area.

But this is only a fraction of the Sun's output. If we could tap directly into its energy, we would have no need to build our own artificial power plants here on Earth.

The Sun's Appearance

We see the Sun shining as a bright yellow-orange ball and looking much the same size as the Moon. The Sun looks a lot closer to us than it really is. This can be best appreciated when we realize that a passenger jet, travelling at over 600 miles per hour, would have to fly continuously for over 17 years to reach the Sun's outer surface.

Light emitted by the Sun is made up of various colors. These mixed together produce white light. When the Earth's atmosphere filters out part of the incoming blue light belonging to the Sun, it makes it appear more orange-red than it really is. The filtered-out blue light is scattered in our atmosphere, and this is why the sky looks blue in daylight.

At sunrise and sunset you will often notice that the Sun looks blood red. This is because at these times the sunlight has to travel through a greater thickness of air, which causes even more filtering and scattering. As the Earth's atmosphere also bends (refracts) incoming light, and as the bending of the light is greatest near the horizon, the shape of the Sun at sunrise and sunset looks distorted. At these times, it often appears oval-shaped, or flattened.

Sunspots

The most interesting visible features of the Sun's surface are dark spots called sunspots.

The very largest ones can sometimes be seen with the naked eye, but most sunspots require some optical aid. When Galileo turned his telescope on the Sun in 1610, he could not believe what he saw and withheld his observations for a year! But sunspots are real enough, and old records from China reveal that they were seen in very early times.

A typical sunspot has two parts. At the center is a very dark area called the *umbra* (Latin for "shade"). This is surrounded by a lighter zone called the *penumbra* (Latin for "almost shade").

How to See Sunspots

In spite of the Sun's great distance, NEVER attempt to look at it *directly*. Looking at the Sun *directly* can cause permanent damage to your eyes and may lead to blindness.

Even special Sun filters sold in optical stores are NOT safe to use, and homemade smoked glass may crack with the heat.

The best way to see sunspots safely is to project the Sun's image through a telescope onto a stiff white card behind the eyepiece as shown in the illustration.This method of viewing sunspots was invented by Galileo in 1610.

A safe method of projecting the image of the Sun to view sunspots.

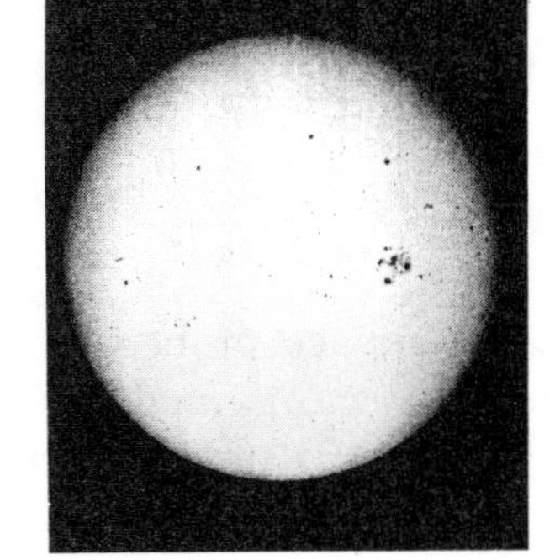

The appearance of the Sun and its spots when projected onto white card by a small telescope.

The center of a sunspot *looks* dark, but it is still a very hot region. It only appears dark in contrast to Sol's brighter areas.

No one really knows *exactly* what sunspots are. We do know that they are very active magnetic "whirlpools" as well as regions which emit tiny subatomic particles that are blown outward at high speeds. Astronomers call these subatomic emissions the solar wind.

The solar wind particles which pass close by the Earth are attracted by its magnetic poles. Inside our atmosphere they trigger off colorful electrical storms known as the aurora.

Because the Sun, like the Earth, spins on its axis, sunspots move across its surface in an east to west direction. As the Sun's surface is not rigid, sunspots travel at different speeds. At its equator the Sun rotates in about 25 days, but towards its poles the period lengthens to 33 days.

In 1843 a German amateur, Heinrich Schwabe, who had observed the Sun for 43 years, discovered that sunspots came and went over a cycle of 11 years. This is now known as the Solar Cycle. The last maximum year for sunspots occurred in 1990.

The Moon

The Moon revolves around the Earth in 29½ days. Its diameter is 2,159 miles, and its distance from the Earth is 238,000 miles. In some ways it is our little twin-sister planet, but there are big differences between us, just as there are big differences between the Earth and the rest of the planets in the solar system.

After the Sun, the Moon is the most influential body in our sky. The pull of the Moon causes much of the tides in the oceans. Compared with the distances of the Sun and the other planets, the Moon is very close to us. It is closer to us than any other body except for straying meteors and meteorites.

The Features of the Moon

Before space probes reached the Moon, all our knowledge of its features was gained through telescopes. After Galileo looked at the Moon in 1610, skywatchers began to draw detailed maps of it.

The dark areas on the Moon, which can be seen with the naked eye, were thought to be seas like those on Earth. They were therefore called *maria* (Latin for "seas"), and they were all given Latin names.

We now know there is no visible water on the Moon's surface, but the traditional names for these features remain unchanged on lunar maps. When the Moon is full, the seas often suggest fanciful figures to the naked eye. Most have heard of "the Man in the Moon." It is fun to spot other less well-known shapes such as "the Woman Bent Over Reading a Book," the very easy "Crab's Claw," "Jacob in the Moon," "the Hare in the Moon," and "the Toad in the Moon."

Generations of young skywatchers have traced out such figures. One, Julius Schmidt, who was born in Germany in 1825, went on to become one of the greatest Moon watchers of all time. As a boy,

The Moon 22 days old, near Last Quarter.

The Moon 7 days old, near First Quarter.

he used to perch his small telescope on the garden gate night after night to watch the Moon and he would tell his father that, when he grew up, he would make it his life's study.

Thirty-four years later he finished his great map of the Moon. It had a diameter of eight feet and was divided into twenty-five sections. It showed 32,856 separate craters, scores of mountain chains, and thousands of smaller features. In the meantime he had been appointed director of the Greek National Observatory in Athens.

It was while in Athens that he was first to spot a new star that suddenly appeared in Cygnus in 1876. Schmidt had a great knowledge of the constellations and this enabled him to discover several variable stars.

All mapping of the Moon's surface is now done by photography via space probes. Geologists have been able to look at rock samples brought back from lunar missions, and these have provided us with a history of our "twin planet."

The Moon's surface has been a largely unchanged world for the past 4,000 million years (4 eons). The Moon itself was probably born at the same time as the Earth, 4½ eons ago. *How* the Earth and the Moon were born remains one of the great mysteries.

The dark maria regions are really old lava plains. Many of the large craters we see were gouged out by asteroids and large meteorites bombarding the lunar surface. Other craters are of volcanic origin, but because the Moon's gravity is only one-sixth of that of the Earth's, the volcanoes that were once active there were much different from those we know on Earth. The jagged mountain ranges we find along the borders of the maria rise to over 23,000 feet.

Exploring the Moon's Surface

A small telescope is the ideal instrument for exploring the Moon, but even binoculars will reveal the largest craters and the more prominent mountain chains.

Whatever you use, do *not* wait until the Moon is full! Although you may think that this is the best time to study the Moon, it is the *wrong* time because at the time of the Full Moon the Sun is shining from directly overhead on the Moon, and the lunar features therefore cast no dark shadows.

A map of the Moon with a key to the principal features.

Craters:
1 Albategnius
2 Archimedes
3 Aristarchus
4 Aristillus
5 Autolycus
6 Copernicus
7 Piccolomini
8 Aristoteles
9 Gassendi
10 Grimaldi
11 Herodotus
12 Kepler
13 Langrenus
14 Vendelinus
15 Plato
16 Posidonius
17 Ptolemaeus
18 Schickard
19 Fracastorius
20 Tycho

Mountain Features:
a Leibnitz Mts.
b Pyrenees Mts.
c Altai Mts.
d Haemus Mts.
e Apennine Mts.
f Caucasus Mts.
g Alpine Valley
h Carpathian Mts.
i Riphaen Mts.
j Cordillera Mts.
k D'Alembert Mts.

Mare Features:
A Mare Australe
B Mare Smyth
C Mare Humboldtianum
D Mare Foecunditatis
E Mare Crisium
F Mare Nectaris
G Mare Tranquillitatis
H Mare Serenitatis
I Mare Vaporum
J Mare Nubium
K Mare Imbrium
L Mare Humorum
M Oceanus Procellarum
N Mare Frigoris
O Sinus Medii
P Sinus Aestuum
Q Lacus Somniorum
R Sinus Iridum
S Sinus Roris
T Palus Somnii

Start looking at the Moon shortly after New Moon, when the Moon only shows itself as a thin crescent. Then follow it day by day as the Moon slowly waxes. Now, you will see how beautifully the advancing line of the lunar sunrise gradually reveals the ragged sinuous outlines of mountains and craters.

Remember, you are in effect looking down at the Moon's surface from a point overhead at great height. With x7 binoculars, the Moon's distance "below you" is about 35,000 miles. With a small telescope magnifying x25, the distance is reduced to less than 10,000 miles.

The Moon's shadows are much darker than those on Earth. This is because the Moon has no atmosphere. Just stepping around a large rock, you pass from dazzling sunlight into midnight blackness.

At the time of the Full Moon, when the mountain chains hide themselves for lack of shadow, you will be able to see all the different maria shown on the map. You will also be able to see the long streaks of light emerging from the 54-mile-diameter crater Tycho, which often suggests a "peeled orange."

Tycho was formed long ago when an asteroid collided with the Moon and the resulting splash of material radiated out and left its mark on the surrounding area. Other large craters, like Copernicus, were formed in a similar fashion.

Around the time of Full Moon, some areas look so bright earlier astronomers thought them to be active volcanoes. One such bright area surrounds the crater Aristarchus. Even in binoculars this is a very bright spot. In small telescopes it is so bright it may dazzle your eyes. For this reason some people use a lunar filter (a neutral filter) when looking at the Moon. A lunar filter can be purchased at a photographic or optical store if the telescope doesn't come with one.

The Phases of the Moon

As the Moon revolves around the Earth in 29½ days, in a counterclockwise direction, it appears to change its shape. The changes in the Moon's shape are called the Phases of the Moon.

When a New Moon occurs, the Moon lies in the direction of the Sun, and we cannot see it because its lit-up face is pointing away from us.

A few days after New Moon, as the Moon moves around the Earth, we begin to see it as a very thin crescent in the western sky shortly after sunset. Although people customarily refer to it at this time as "the New Moon," it is by now several days old.

Also at this time we may see the dark unlit part of the Moon faintly illuminated by sunlight reflected off the Earth. This is called earthshine. In the old days, country folk used to say that this was the Old Moon cradled in the arms of the New Moon.

As the Moon continues on its month-long journey around the Earth, the crescent shape grows larger, or waxes, so we gradually see more of the Moon's sunlit face. When we see *half* the Moon's sunlit face, this is called the Moon's First Quarter. It may seem strange to you to refer to half the Moon as "a quarter," but in terms of the whole Moon, the lit-up part, *visible to us on Earth,* is only a quarter.

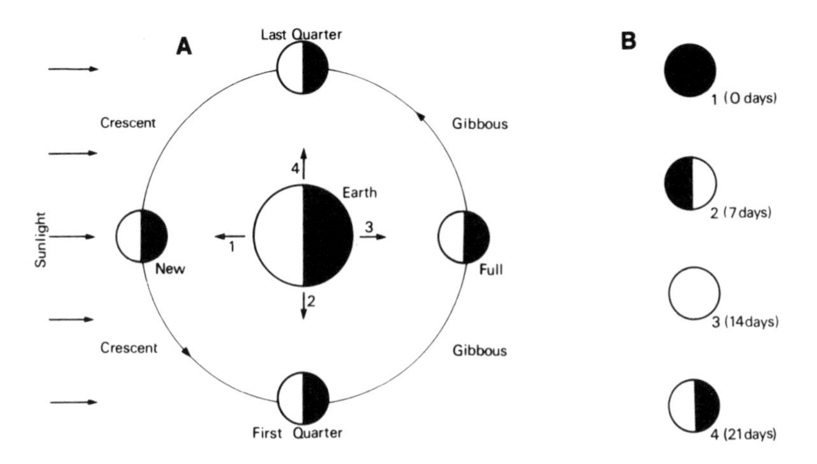

The phases of the Moon (**A**) and the appearance of the Moon at New, First Quarter, Full, and Last Quarter (**B**).

About seven days later we see the whole of the Moon's sunlit face, and this is called the Full Moon.

After Full Moon, as the Moon travels on, its sunlit face (visible to us) begins to shrink, or wane. When it wanes to exactly half again, this is called the Last Quarter. From Last Quarter, as it creeps closer to the Sun, it will become crescent-shaped again until it finally disappears from our view in the dawn sky near the time of the next New Moon.

The Moon always keeps its same face pointing towards us because it is locked into the Earth's gravitational pull. As a result, we did not know what the other side of the Moon looked like until space probes were sent to fly around it.

Eclipses

Eclipses are events ideal for viewing either with the naked eye or with binoculars. There are two kinds: eclipses of the Sun (solar eclipses) and eclipses of the Moon (lunar eclipses). Some eclipses are called partial eclipses—when only part of the Sun or Moon is covered; other eclipses are called total eclipses—when all of the Sun or Moon is covered.

Total solar eclipses happen because of the remarkable coincidence that the Moon and the Sun look to be the same size. The Moon, of course, is much smaller than the Sun. It only appears as large because the Moon is much closer to us.

A total eclipse of the Sun in Siberia in 1968.

Why Solar Eclipses Occur

Eclipses of the Sun only occur when the Moon comes *exactly* in line between the Earth and the Sun. Because the Moon's orbit around the Earth is oval shaped, or elliptical, the Moon is not always the same distance from the Earth. Therefore, from the Earth, the Moon sometimes appears slightly smaller than the Sun and at other times slightly larger. Because of this we can see different kinds of eclipses.

You might expect eclipses to occur every New Moon when the Moon is then in line with the Sun. Eclipses do not happen every month because the Moon's orbit around the Earth is slightly tilted. Eclipses of the Sun only occur when the Moon passes a crossing point (called a node) in its orbit at the same time as the New Moon. If at this time the Moon is larger than the Sun, it will cover it completely and so bring about a total eclipse of the Sun.

Sometimes, however, the apparent size of the Moon will be smaller than the Sun. This will reveal a narrow width of Sun all

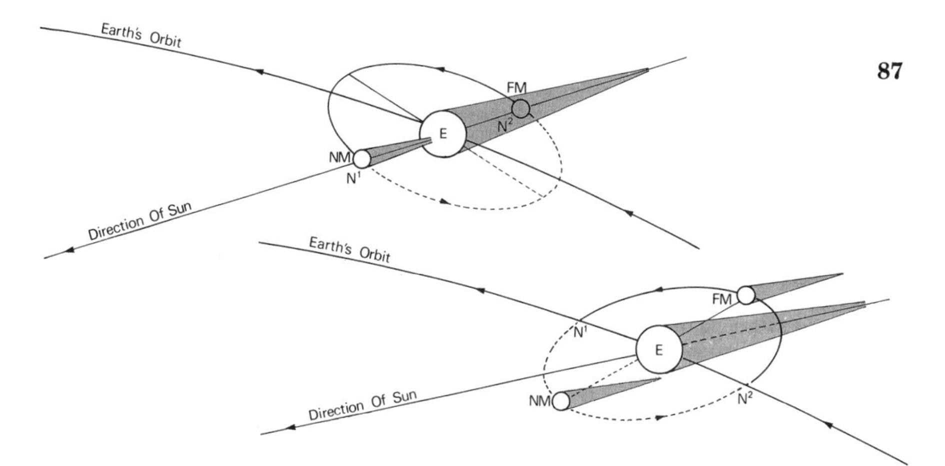

Eclipse occurrences. Owing to the Moon's orbit being inclined a little over 5°, eclipses can only occur at New Moon (NM) and Full Moon (FM) when the line of nodes (N^1–N^2) of the Moon's orbit coincides with the direction of the Sun.

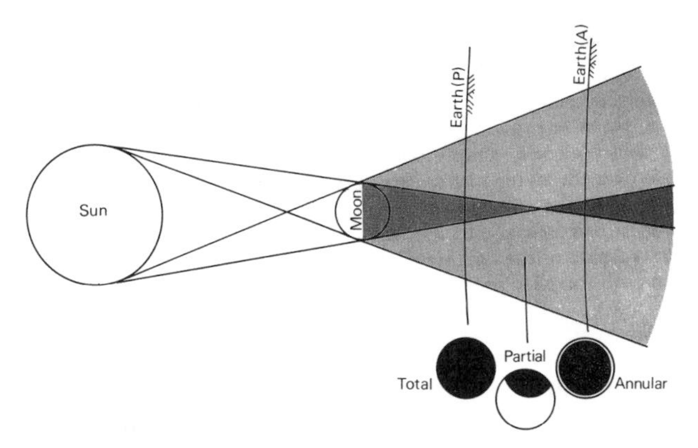

Solar eclipses can be total, annular, or partial, depending on how the Sun, Moon, and Earth are lined up. When the Moon and Earth are closest, at perigee (P), a total eclipse occurs. When the Moon is more distant, at apogee (A), annular eclipses occur. When the Sun, Moon, and Earth are not exactly in line, partial eclipses occur.

around the Moon. Such an event is called an annular eclipse of the Sun (annular is Latin for "ringlike").

Sometimes the Moon will not be quite in line with the Sun and cover only a portion of it. This is called a partial eclipse of the Sun.

Why Lunar Eclipses Occur

Eclipses of the Moon occur when the Earth comes in line between the Moon and the Sun at the time of the Full Moon. This causes the Earth's shadow to pass over the Moon.

As the Earth's shadow creeps slowly across the face of the Moon, it has a curved edge to it. This was proof to the early Greeks that the Earth was a round body.

Like solar eclipses, lunar eclipses are sometimes only partial ones—when the Sun, Earth, and Moon are not exactly in line.

Finding Out When Eclipses Occur

Total solar eclipses are rare events for any one area. The reason for this is that the Moon casts its shadow over a very narrow track on the Earth's surface.

Partial eclipses of the Sun and total and partial eclipses of the Moon are much more common. At least one of these events is bound to occur soon in your area. Be prepared!

Newspapers are a ready source of information for when eclipses are due to take place, but these sources usually announce them only a day or so in advance. The best way to prepare for one is to look in an astronomical yearbook or in a popular astronomical magazine at your local library. In these publications you will find dates and times, and they will tell you what you may expect. If in doubt, talk to your local librarian.

Eclipse Events

Total solar eclipses are much more impressive spectacles than annular or partial eclipses of the Sun. At the time of totality the whole sky grows dark, and the brighter planets and stars become visible. Birds, thinking that the Sun has set and night is falling, fly home to their nests. However, totality in one spot lasts only a few minutes. The longest possible time in any one spot is a little under 8 minutes.

An annular eclipse may last 12 minutes; a total eclipse of the Moon may last up to 4 hours; partial eclipses of the Sun and Moon vary, depending on how much is eclipsed, but they will usually be longer than half an hour.

Because total eclipses of the Sun are fairly rare events for any one spot, some astronomers travel around the world to see them. In 1968, I travelled all the way to Siberia to watch one that lasted only 38 seconds! But the experience was worth it. Astronomers study eclipses to find out more about the Sun's outer layers. Sometimes at the time of totality, when the sky is darkened, new spectacular comets are discovered.

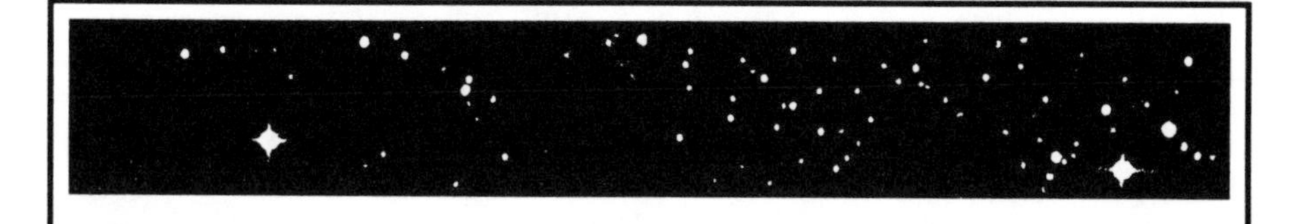

Planet Watching

Mercury

Mercury is the closest known planet to the Sun. As seen from Earth, it is never far away from the Sun. Because of this, finding Mercury can be quite a challenge. It can be very difficult to see it if you live in a town or city where tall buildings and pollution obscure the horizon, but with perseverance it *can* occasionally be spotted even there. As a youth I lived on the outskirts of a large industrial city, but I regularly found Mercury *once I discovered where to look.*

The best times to search are just *after* sunset or *before* sunrise when Mercury is at one of its elongations from the Sun. It will never be more than 26° away—about fifty times the Sun's diameter. At these times it can be seen with the naked eye, but better with binoculars. It will then be seen shining like a bright star of mag –1.8.

In the morning sky, at Mercury's westward elongation, when it is at its brightest, it has often been seen by people not specifically looking for it. Even in the evening sky, it can be noticeably bright. A famous American skywatcher, Garrett P. Serviss, who lived about a hundred years ago, once poetically described it as "glittering like a globule of shining metal through the fading curtain of a winter sunset."

With a small telescope we cannot see any of Mercury's surface features. What we can see are its phases. Because Mercury is an inferior planet (inferior meaning, in this case, that it lies *inside* the Earth's orbit and closer to the Sun), it shows the same changes in shape as the Moon does as it travels around the Earth. As Mercury goes around the Sun in only 88 days, a few days of observation will show these changes.

Until the Mariner 10 probe in the 1970s, little was known about Mercury's surface. Some suspected it was cratered like our Moon, but no one could be sure. Mariner confirmed that Mercury is

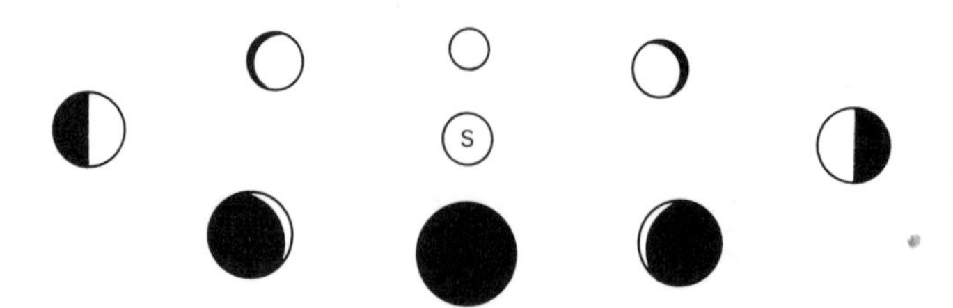

Mercury and Venus are called inferior planets. This is not because of their size, but because they are closer to the Sun than the Earth. As a result, from the Earth, Mercury and Venus show phases as they orbit the Sun.

mountainous and very heavily cratered. It is an arid and airless world, totally lacking in life.

Mercury has a diameter of 3,033 miles and rotates in 58.65 days. Its orbit is very eccentric and this carries it to within 28½ million miles of the Sun's surface at its closest and 43½ million miles at its farthest.

At its closest, Mercury's surface bakes in a temperature of 800° F. As no living thing could survive there for long, it is believed that Mercury is a very hostile world where life has never existed.

Venus

No one could ever accuse Venus of being an elusive planet. From time to time it is so bright that it attracts the attention of people with no knowledge of skywatching. As a result, it is the most frequently misidentified object in the heavens, often the target of UFO spotters.

After Mercury, Venus is the next planet outward from the Sun. Because it also lies inside the Earth's orbit, it is an inferior planet and, like Mercury, shows phases. These phases were first seen by Galileo in 1610. They proved to him that Venus, like all the other planets, revolved around the Sun.

Venus lies at a mean distance from the Sun of 67.2 million miles. It revolves around the Sun in 224.7 days and turns on its axis in 243.16 days. Venus is approximately the size of the Earth, having a diameter of 7,523 miles.

Venus is sometimes so bright, shining at mag –4.4, that we can see it with the naked eye even before the Sun has set. Around this time we can even see it with the naked eye at midday if we know *exactly* where to look. After dark it can be so bright it casts a shadow during its brightest phases.

People with keen eyesight often claim that they can see the phases of Venus with the naked eye, but this is doubtful. Whatever, a pair of ordinary binoculars will readily show them, making Venus appear like a tiny version of our own Moon. Even with a telescope there is little else to be seen on Venus because the planet's surface is permanently shrouded in thick cloud.

Before the space age, people could only guess what lay under Venus's thick clouds. Some believed that life flourished there. Alas, the various probes have now unveiled a very fiery Venus and shown it as a world hostile to life as we know it.

Beneath its thick clouds the surface is scattered with craters. Some of these were formed by huge volcanoes and others by colliding asteroids. Rugged mountains tower 6 to 9½ miles high—much higher than Mount Everest. Some of the volcanic features may still be active.

Venus's surface temperature is nearly 1,000° F—so hot that lead melts. Its atmosphere is so dense it has a pressure of 1,260 pounds per square inch, compared with a mere 14 pounds on Earth. Any visiting astronaut, unless properly protected, would be instantly burnt to a crisp and crushed to death. From what we have so far learned about Venus, it is doubtful that life has ever existed there.

Mars

Mars is the next planet outward from the Earth–Moon system and a favorite object among skywatchers. For a long time it represented a mysterious, unknown world where intelligent life could exist. Everyone had heard of the canals of Mars and the green, bug-eyed monsters that inhabited the planet.

Science-fiction writers like Edgar Rice Burroughs and H. G. Wells, and others who followed in their footsteps, spun fantastic tales about the supposed Martians. While no one really believed such creatures existed, it was not until the 1960s and 1970s that we could be absolutely sure what conditions were like on the surface of Mars.

Some believed that the reported canal-like features were actual waterways constructed by Martians to allow water to flow from the planet's polar caps to irrigate the hot deserts around its middle regions. While the polar caps were plain enough, not everyone could see the supposed canals.

These canals were discovered in 1877, by an Italian astronomer called Giovanni Schiaparelli, when Mars made a close approach to the Earth as it does from time to time. Schiaparelli announced that he had noticed a network of fine lines crisscrossing the planet's surface. He called them *canali* without implying in any way that they were artificial features of the Martian landscape. In translation, however, the word canal was used, and with it the implication that they had been constructed by some intelligent life.

Soon after, others claimed to have seen Schiaparelli's canals. A rich American called Percival Lowell built an observatory at Flagstaff in Arizona to make a special study of them. It was Lowell who furthered the idea (which Schiaparelli had never intended) that the canals and some dark, round spots connecting them, which Lowell called oases, were undoubtedly intelligent creations.

Many people disagreed with Lowell, but others were active supporters of the canal theory. More and more observers began to see them, even people looking at Mars with small telescopes. However, not all who claimed to have seen the canals thought them artificial but believed they were definite surface features.

There the matter rested until 1965 when *Mariner 4* passed within 8,700 miles of Mars and sent the first photographic close-ups back to Earth.

The photographs showed no canals and no oases, but to everyone's surprise lots of craters that resembled those on the Moon. Some were not surprised about the craters, for the great American skywatcher E. E. Barnard had thought he had seen crater-like features on Mars many years before. It is probable that Lowell's "oases" were, in fact, large craters seen at the limit of visibility. But of the canals themselves, there was no trace.

Later probes confirmed that fiery volcanoes had once burned over wide areas of Mars and that large tracts of the planet experience violent dust storms which rage for months on end. There are huge valleys and wide cracks in the arid surface, but no sign of a living thing. The age-old myth that Mars may be an inhabited planet was shown to be only science-fiction writers' make-believe.

Mars is a smaller planet than the Earth, with a diameter of 4,200 miles. It lies at a mean distance from the Sun of 142 million

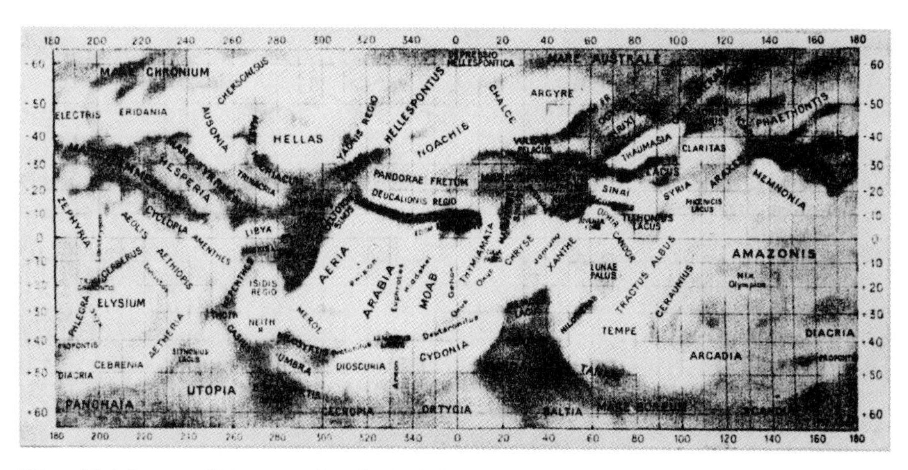

The official map of Mars naming (in Latin) its various surface features. Some of these features can be seen in small telescopes when Mars is close to the Earth.

miles and takes 1.88 years to go around the Sun once.

Mars comes close to the Earth about every two years, but closer approaches come at 15- to 17-year intervals. At its closest it shines like a very bright orange-red star of mag –2.3. At these times no one has the slightest difficulty in recognizing it.

The red color of Mars attracted much attention in the ancient world and earned Mars its name as the mythological Roman God of War.

In a small telescope, when Mars is farthest away, it shows up only as a tiny disc. But at its closest (in opposition, as it is called), a small telescope with an eyepiece of x75 will enlarge it to the apparent size of the Full Moon.

At the time of opposition, very small telescopes will often show some subtle shadings on Mars. Even a 1-inch telescope will do this, but a minimum magnification of at least x50 is necessary.

With a 2-inch telescope and larger magnification, more of the darker areas of Mars become visible. These, like the darker areas on the Moon, were once thought to be seas. A 2-inch telescope will also show one of the polar caps, but which pole you see will depend on the hemisphere that is tilted towards the Earth at the time of viewing. These polar caps are formed from a mixture of water-ice and carbon dioxide and they wax and wane according to the Martian seasons.

A 4- to 6-inch telescope will show many more features. If you observe over a longer period, you will notice these features ever changing as Mars rotates in 24 hours, 37 minutes, and 23 seconds.

Mars has two small moons called Phobos and Deimos, but these are very faint and quite beyond the reach of ordinary telescopes.

Jupiter

Jupiter runs Mars a close second as a favorite planet with skywatchers. Even with ordinary binoculars you can see Jupiter as a small, yellowish oval disc with its brightest moons strung out alongside, glittering like tiny pearls.

How many of its brightest moons you will see at any one time will depend on where each moon is in its orbit around Jupiter. Sometimes all four will be bunched to one side; at other times they will be divided either side of the planet; sometimes one or more will be passing in front of Jupiter (in transit); or one may be hidden behind (occulted by) the planet.

The four brightest moons are sometimes called the Galilean satellites (or moons) as they were discovered by Galileo in 1610. Otherwise, they are called by their names: Io, Europa, Ganymede, and Callisto.

With ordinary binoculars you will not be able to see Jupiter's cloud belts, but a 2-inch refracting telescope will reveal two or more of them. With a 3- or 4-inch telescope you will be able to see quite a lot of fine detail in the various cloud belts.

Jupiter's cloud belts and the famous Red Spot are the dominant features of the planet. When Jupiter is closest to us (in opposition), a magnification of only x40 will show it as large as the Full Moon.

Jupiter lies beyond the orbit of Mars and takes 11.86 years to orbit the Sun at a mean distance of 483.3 million miles.

Among the planets it is the oval giant, measuring 89,424 miles across the equator and 83,000 miles from pole to pole. Jupiter spins on its axis in 9 hours, 50 minutes, and 30 seconds; this is faster than any other of the major planets.

Jupiter is a giant planet *two-and-a-half times* bigger than all the other planets put together. A lot of its bulk is made up of dense gases rather than solid material. Some astronomers believe that if Jupiter had been just a little larger, it would have developed into a star instead of a planet.

Most of our information about Jupiter, and its many moons, comes to us from the *Pioneer* and *Voyager* space missions of the 1970s.

It was found that Jupiter's thick atmosphere is made up of hydrogen and helium gases and the upper cloud belts are formed from small droplets of ammonia. The prominent Red Spot feature is a great semi-permanent storm-center that rages inside Jupiter's atmosphere.

One surprise discovery from the missions was the existence of a thin, dark ring surrounding the planet. Invisible from Earth, this tenuous will-o'-the-wisp feature is quite unlike the solid-looking rings of Saturn.

The biggest surprises were found on the four biggest moons. Callisto is covered with craters and ice. Ganymede, which in size is larger than the planet Mercury, is also icy and scored by cracks and strange, long grooves and furrows. Neither of these two moons appear to be active worlds. In contrast to these, Europa appears quite different—where something strange *might* still be happening. In photographs its surface looks like a cracked egg-shell.

The last of the Big Four satellites, Io, which is a little larger than our Moon, held the biggest surprises. When visited by *Voyager*, astronomers at Mission Control at first wondered if they were gazing at a pizza! Io's sulphur-coated surface was scattered with red, white, yellow, brown, and black patches. Never before had the astronomers been greeted by such a bizarre sight. They quickly checked to see if the mission cameras had gone haywire.

Io was no dead museum piece. On its surface, things were definitely happening! Astronomers gazed fascinated at what was taking place. There were eight very active volcanic plumes of liquid sulphur, and material was being thrown up hundreds of miles above Io's surface. The Earth was no longer the only active body in the solar system!

Before the missions, Jupiter was known to have at least twelve moons. The missions discovered another four. Most of Jupiter's moons are tiny, icy worlds only a few miles in diameter. It is the four largest that attract the most attention. All are very curious places. Europa, in particular, might prove to be a very unique place, and astronomers look forward to more detailed explorations of it when the Galileo Mission reaches Jupiter in 1995.

Saturn

Because of its unique ring system, Saturn has long been a showpiece in the sky. When the rings are at their widest, they can be seen with the smallest of telescopes.

Saturn, in size, is second only to Jupiter and, like Jupiter, is noticeably flattened at its poles. Its equatorial diameter is 74,914 miles, and its polar diameter 66,400 miles. Saturn takes 29.46 years to orbit the Sun at a mean distance of 886 million miles. Like Jupiter it rotates rapidly, but Saturn takes one hour and ten minutes longer, giving it a period of 10 hours 39 minutes.

Saturn's rings are not really within range of binoculars. However, when the rings are at their widest, even in binoculars there is a suspicion that the planet's image is oval. A modern 2-inch refracting telescope with a x25 eyepiece will show the rings much more clearly than Galileo ever saw them when he looked at Saturn in 1610. The lenses in Galileo's telescopes were imperfect, and looking at the planet he thought he saw three globes, a large one with a smaller one attached on either side of it. This suggested that Saturn was made up of three bodies in a row.

As Galileo continued to watch Saturn over the next two years, he noticed that the smaller bodies on either side gradually disappeared, and by 1612 they were invisible! He now doubted his own observations and stopped looking at the planet.

Poor Galileo. His telescopes were not quite good enough to show him what was really happening around Saturn. What Galileo had seen on either side of the planet were the curved sections of the ring closing up. Because Saturn's rings are always tilted in the same direction, they *appear* to open and close from where we see them back on Earth. This happens over a period of several years as Saturn and Earth go around the Sun in their orbits. Sometimes the rings are edge-on to the Earth and because they are very thin, they disappear from view for a time. The rings were edge-on in 1612 when they disappeared in Galileo's telescopes.

The rings were first properly identified in 1656 by the Dutch astronomer Christiaan Huygens, who also discovered Saturn's brightest moon, Titan. This moon (mag 8.4) can be seen with

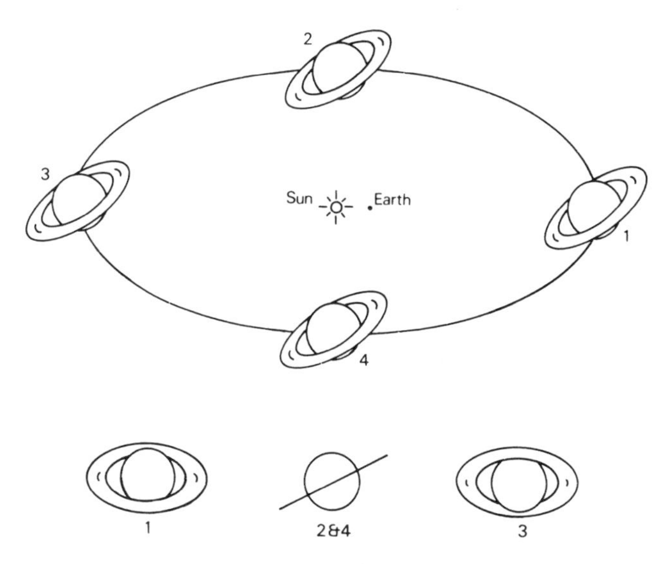

The changing appearance of Saturn's rings as seen from Earth. The rings are at their widest at 1 and 3. In position 2 and 4 the rings, seen edge on, appear to close up.

binoculars. Three more satellites, Tethys, Dione, and Rhea, are visible in a 3-inch telescope, and with a quality 4-inch telescope two more, Iapetus and Enceladus, come into range.

Saturn's family of satellites number seventeen, but most of them are faint objects beyond the range of ordinary telescopes.

Well within range of smaller telescopes is a dark division in the ring system called the Cassini Division, so named after the French-Italian astronomer Giovanni Cassini, who first noticed it in 1675. When the rings are reasonably wide open, it can be seen with a 2-inch telescope and a magnification as little as x25. Astronomers coming after Cassini found other divisions in the ring, but these require larger instruments.

The different sections of the ring around Saturn are all made up of myriads of tiny bodies in orbit. Because we view them at a great distance, the rings appear solid. It was the ring system that attracted the special attention of astronomers in charge of the *Voyager* missions, the first of which reached Saturn in November of 1980.

The rings were found to be a lot more complicated than ever

imagined. At close range they were seen to be divided into very intricate patterns, quite invisible when viewed in telescopes from Earth. There were so many separate parts to the rings that one scientist viewing them remarked that he was reminded of the grooves in a phonograph record.

Yet, on final count, they were even more numerous than this, for *thousands* of separate rings were counted. Even the 2,500-mile-wide, dark Cassini division, once believed to be "empty" of jostling particles, had rings inside it.

From the *Voyager* missions, astronomers have concluded that the particles forming the rings range in size from pebbles to small boulders. All are frozen to a temperature lower than –290° F. The thickness of the rings is probably no more than 500 feet, or even less. It is little wonder that Galileo lost sight of them when they were edge-on to the Earth in 1612.

In many ways the rings remain something of a mystery, and astronomers will continue to study the data from the *Voyager* missions for years to come.

The *Voyager* missions confirmed that Saturn was a gaseous body like Jupiter and that storms rage through its very turbulent atmosphere in a similar manner. One storm was clocked at a howling 900 mph, which shades into insignificance the 100- to 150-mph hurricanes we experience on Earth.

Saturn's large family of moons proved to be fascinating and curious worlds like those of Jupiter. Most are icy and cratered. The brightest, Titan, appears like an orange but hides its mysteries under a thick atmospheric blanket. Because of this atmosphere, some astronomers have fancied Titan to be another place in the solar system where life may be found. However, the *Voyager* team measured its temperature at –290° F. This sounds like a very cold habitat for the wellsprings of life to flourish. But who knows?

Uranus

Uranus was the first of the planets to be discovered with a telescope. It was found accidentally on March 13, 1781, by the amateur skywatcher William Herschel while he was examining part of the sky with a homemade 7-inch reflecting telescope.

This German-born amateur is the father figure to all skywatchers. After travelling to England, he showed how a person of humble origins could, by self-education and application, find fame and fortune in the study of the heavens.

Uranus is another member of the family of giant planets, often referred to as the Jovian planets—so named after Jupiter, the largest of them all. Uranus has a diameter of 31,770 miles and lies at a mean distance from the Sun of 1,783 million miles. It takes 84.01 years to go around the Sun and rotates in 17 hours, 14 minutes.

While usually considered a strictly telescopic object, Uranus, at its brightest, can occasionally be glimpsed with the naked eye. At this time, it shines like a very dull star of magnitude 6, but at the very limit of visibility so it needs a very dark sky to be able to see it.

As a telescopic object, Uranus has little to offer the amateur skywatcher, and it really needs a 6- to 12-inch telescope to show its tiny, greenish disc properly.

In many ways, Uranus is a peculiar planet. Its polar regions are tilted over 98° to where you would normally expect to find the equatorial regions. An even more surprising feature of Uranus is the modern discovery that it has a ring system. This was found in 1977 by astronomers looking from Earth and using a special method. Until 1977 it was believed that Saturn was the only planet with a ring system. Since then several planets have been found to have rings, but they are all much fainter features than Saturn's.

Before the *Voyager 2* mission reached Uranus in 1985/86, very little was known about this strange world. Five moons were known, but the *Voyager* cameras added another ten. All of these were named by astronomers after well-known characters in Shakespeare's plays. These days astronomers can have a lot of fun naming new discoveries. Some craters on one of the new moons found by *Voyager* were called: Bogle, Lab, and Butz.

Miranda, one of the previously known moons, had the most startling appearance. Some astronomers believe that this tiny 293-mile mini-world was shattered in the past by an impact with another cosmic body but since then has reformed itself from the leftover pieces!

There are many strange worlds, indeed, at the outer reaches of the solar system.

Neptune

The telescopic discovery of Neptune in 1846 was one of the greatest feats in astronomy. Finding this planet was not the result of an accidental observation, as Herschel's discovery of Uranus had been, but the result of painstaking calculation (see box).

Neptune lies out beyond Uranus at a mean distance from the Sun of 2,793 million miles. This is over thirty times greater than the Earth's distance from the Sun. Neptune takes 164.8 years to orbit the Sun, so it has not yet made one complete revolution since it was discovered.

Neptune has a diameter of 31,410 miles and rotates in 16 hours, 3 minutes. In size, makeup, and rotation period, it very much resembles its neighbor Uranus, and it is the outermost member of the four Jovian planets.

Very little was known about Neptune until it was visited by the *Voyager 2* mission in 1989. In telescopes it was seen to be a bluish, cloud-covered planet accompanied by two faint moons. *Voyager* immediately discovered six more moons.

As *Voyager* approached Neptune, the astronomers at Mission Control were on the lookout for a ring encompassing the planet. Observations from Earth, like those which had detected a ring around Uranus in 1977, indicated that this was a possibility.

Sure enough, *Voyager* detected two bright rings, plus a fainter one. There may even be fainter ones awaiting discovery in the future.

Voyager also detected a large, dark spot on Neptune's cloud-covered, blue surface—not surprisingly, the astronomers named it the Great Dark Spot. Another dark spot was also seen drifting in the clouds, plus a feature nicknamed "Scooter" because it rotated quickly around the planet.

Neptune, as with all the cloud-covered Jovian planets, was found to have strong winds raging through its lower atmosphere. Some blew at speeds of 700mph, turning the planet into a truly inhospitable world. It seems that Neptune is a very cold, lifeless planet made up chiefly of hydrogen gas. Triton, its largest moon, proved to be the coldest place ever visited during all the various *Voyager* missions when a record low temperature of –400° F was recorded.

The Discovery of Neptune

The thrilling discovery of Neptune in 1846 began when two men, quite independently, set out to show in their calculations how some peculiarities in the orbital path of Uranus, which had long puzzled astronomers, might be explained by the pull exerted on it by the presence of a large, unseen planet lying beyond it.

These two mathematicians were the Frenchman Urbain Leverrier and the Englishman John Couch Adams. As both Leverrier and Adams were concerned with theory and were not practising skywatchers, they had to rely on other people to make the search.

Leverrier sent his calculated position of the supposed planet to Johann Gottfried Galle at the Berlin Observatory. He did this because Leverrier had found his own countrymen slow to take up this challenge. The observatory at Berlin had a fine 9-inch refractor, and Galle was known to Leverrier as a very enthusiastic observer. However, Galle was still only an assistant at the observatory and needed permission from the director, Johann Franz Encke, to look for the new planet.

Meanwhile in England, James Challis, professor of astronomy at Cambridge, had started a search from his own observatory some time before, using the prediction made by Adams. The observatory at Cambridge had a fine 12-inch refractor more powerful than the Berlin telescope. Yet, in the end, there was one key factor that was to count more decisively than telescopic power.

Before Leverrier's request arrived in Berlin, Encke heard rumors about a supposed new planet lying beyond Uranus. These rumors had been about for several years, so Encke showed no surprise when Galle showed him Leverrier's letter. Moreover, he did not believe that such a planet might exist. He told Galle, his assistant, that the search for such an object would be a waste of time and withheld his permission to begin the search.

Galle was crestfallen. All day long, after receiving the letter in the morning's post, he badgered his director to change his mind. The truth is that Encke, while a brilliant observer himself, was a little jealous of his assistant's prowess as an observer. However, by late afternoon, tired of being harassed by Galle, Encke finally relented. Galle could search for the supposed new planet that very night if he so wished. As it happened, Encke had a social engagement and would be absent from the observatory.

A young student at the observatory, Heinrich Louis d'Arrest, had overheard the fractious exchanges between Encke and Galle. When permission was granted to Galle, d'Arrest begged him to be allowed to help him in the search. Galle later wrote: "It would have been unkind to refuse the wishes of this zealous young astronomer." With a wry smile he told d'Arrest to be back at the observatory soon after dark. D'Arrest needed no prompting.

It was now that fate interceded to help the astronomers. By good fortune, a new star chart covering the very area they were to search that night had been received just a few days before.

Dusk had barely fallen when Galle took his place at the telescope. Then, when the sky had darkened enough, he began to call out the various stars as they passed into his view in the eyepiece. D'Arrest, seated at a desk nearby with the chart spread out in front of him, ticked off each one in turn.

Only a few minutes had passed before the young student suddenly caught his breath. He called out: "That last star is not on the map!"

Had the new planet been found so easily?

The "star" was not on the map, but neither Galle nor d'Arrest was yet sure it was the planet. Even in their fine 9-inch telescope, the tiny disc of the object they had found was barely discernible. Could it be something else?

By midnight Encke was back at the observatory, having left his social function after an urgent message arrived informing him of events. He was very cautious about what object it might be. He did not want to be made a laughingstock in the astronomical world by announcing the discovery of a new planet from *his* observatory before *he* was sure that it *was* a new planet.

It was not until the following night when the strange "star" was seen to have shifted position from the previous night that Encke agreed to announce to the world that a new planet had been found.

But this is not quite the end of the story. Later it was shown that Challis in Cambridge had actually seen the new planet weeks before the Berlin astronomers. However, even with his larger telescope, he had failed to recognize it as an interloper. He had simply marked it down as a star.

The trouble was that Challis, unlike the Berlin astronomers, did not have a chart for that part of the sky. He therefore set out to make one of his own, hoping to spot a moving object during the course of his charting. Looking at one particular "star," he remarked to his assistant: "It seems to have a disc," but instead of switching to a higher magnification right away, he decided to look at it again the following night. The delay was fatal, for the following night "the Moon was in the way." Thus he let slip from his grasp the greatest astronomical prize of the nineteenth century.

Pluto and Chiron

Pluto is the most distant of the known planets and is in many respects one of the oddest. It was discovered in 1930 by Clyde Tombaugh at the Lowell Observatory in Arizona during a photographic search of the sky for a new planet lying beyond Neptune.

Percival Lowell, who had founded the observatory, had calculated, like Leverrier and Adams had done earlier with Uranus/ Neptune, that another planet lay beyond the orbit of Neptune. The task of trying to locate this unseen planet had fallen to Tombaugh, a one-time farm boy who sought fame among the stars.

Because Pluto is such a small planet and could have little influence on the path of Neptune, most astronomers now agree that this discovery was more an accident than anything else.

Pluto lies at a mean distance from the Sun of 3,666 million miles and takes 248 years to orbit once. Its diameter is only 1,519 miles, and it rotates in 6 days, 9 hours.

Because Pluto's orbit is egg-shaped, it sometimes lies closer to the Sun than Neptune does. Since 1969, and for several years into the future, Pluto will be travelling inside Neptune's orbit.

In 1977 a moon of Pluto's was discovered and named Charon. Its diameter is 745 miles, which is half that of Pluto's. The two objects are separated by only 11,800 miles. They are locked together in a strong and strange gravitational embrace, so they revolve around a common center of motion like an unequal dumbbell.

Pluto is not an object for ordinary skywatchers using small telescopes. It is a faint mag 14.5 object and is very elusive, even for experienced visual observers armed with large telescopes. To have any chance of glimpsing it you require a 15- to 18-inch telescope and you need to watch over several nights to see its tiny image shift across the background stars. It is so elusive and faint, I have only seen Pluto in real life, rather than in a photograph, three times in my observing career.

An object of similar kind, which is even stranger than Pluto, is Chiron (not to be confused with Charon). Chiron was discovered in 1977 and moves in an orbit between Saturn and Uranus. After its discovery it was given the status of an asteroid, being no more than about 125 miles in diameter. Nevertheless, in the 1980s it developed a nebulous envelope like a comet. But whether it is a comet or an asteroid, or a body halfway between the two, is still uncertain.

Between the Planets

Asteroids, or Minor Planets

The ancients knew of five planets, and it was only after Herschel discovered Uranus in 1781 that astronomers began to wonder if there were others.

Looking at the planetary orbits, they had noticed a curious wide gap between those of Mars and Jupiter, compared with the gaps separating the other planets. Was this a region that held a missing planet?

Around the year 1800, a group of enthusiastic German skywatchers banded together to form a society to track down the supposed missing planet. They became known as the "Lilienthal Detectives." Their task was to investigate the narrow strip of sky either side and along the ecliptic (the zodiacal zone where the planets move), dividing it into twenty-four equal search zones, one per skywatcher detective.

By a curious coincidence, a new planet was spotted on January 1, 1801, by the Italian Giuseppe Piazzi from Palermo in Sicily, who was not included among the Lilienthal Detectives and before the search had really started. Piazzi was not even searching for a new planet but was trying to make a catalogue of stars. His discovery, like Herschel's discovery of Uranus, was one of pure chance.

Then Piazzi lost sight of the new planet. However, one of the detectives, a Baron von Zach, was now alerted. With the help of an orbit computed for the new planet by a brilliant young German mathematician, Karl Friedrich Gauss, von Zach finally caught up with it almost a year later on December 31, 1801.

There were some surprises in store for the detectives. The object discovered by Piazzi, later named Ceres, was indeed a planet, but only a very small one—no bigger than 600 miles in diameter. The missing planet was not the kind the detectives had been looking for. Were there others?

Another of the detectives, Heinrich Olbers—by day a skilled practising doctor and by night a dedicated skywatcher—had become very familiar with the fainter background stars in Virgo. On March 28 he noted a mag 7 star that looked very unfamiliar to him. Suspicious that it might be another planet, he tracked it carefully and noted that it moved. Indeed, Dr. Olbers had discovered a second mini-planet.

It was later named Pallas and proved to be even smaller than Ceres, measuring only 110 miles in diameter.

In the following years, many minor planets, or asteroids as some people prefer to call them, were found. At the present time, scores of new ones are added to the list every year. Altogether they probably number at least 30,000—perhaps even more.

Most asteroids orbit the Sun between Mars and Jupiter, but others stray *inside* the Earth's orbit, and from time to time we experience "near misses" with these wandering bodies. None, however, within modern times, has come closer than the Moon, but in the distant past several have hit the Earth and gouged out large craters. One such close encounter is suggested to have killed off the dinosaurs about 60 million years ago.

Since their discovery, some enthusiastic skywatchers have spent much of their lives hunting new asteroids. Hermann Goldschmidt, who lived in Paris, was a famous historical artist by day and a dedicated asteroid hunter by night. He is said to have spent every clear night sweeping the sky from the window of his studio attic in the heart of Paris, quite oblivious to the rowdy nightlife going on in the streets below him. He discovered fourteen new asteroids, using for much of the time a very small telescope about the same size as that used by beginners today.

With photography, professional astronomers have discovered hundreds of asteroids. Karl Reinmuth, a German astronomer, discovered 246 asteroids, and another German, Max Wolf, 232. Most of these objects are very small and only a few miles in diameter. Gravity is so weak on these asteroids that you could jump right off into space.

Now astronomers can choose the name of any new asteroids they

There is strong evidence to suggest that there is a family of more distant asteroids. This evidence is provided by the discovery in 1991 of an asteroid which has an orbit stretching from Mars to out beyond Uranus. This object, designated 1991 DA*, takes 42 years to orbit the Sun and is probably no bigger than 5 miles across.

In 1992 an even more distant one was found. This was designated 1992 AD. At its closest to the Sun it lies inside Saturn's orbit, but at its farthest it lies out beyond Neptune. It takes 93 years to make one orbit of the Sun, and is probably at least 100 miles across.

Both these asteroids may be bodies like Chiron (see above), halfway between comets and asteroids. However, they are so faint that none of their physical details have yet been deciphered. What they *really are* must, for the present, remain a mystery.

In early September 1992, a very exciting discovery was made by two astronomers in Hawaii who were searching for a possible tenth planet. The new body, designated 1992 QB1, was found to be a small, icy object about 130 miles in diameter lying beyond Pluto's orbit and is possibly some kind of cometary body like Chiron. This object may prove to be the first discovery of a whole new family of very distant mini-bodies lying at the frozen fringes of the solar system.

* Provisional asteroid designations are given at the time of discovery, i.e. 1991, 1992, etc. The letters which follow the year are a code devised to identify the different ones in order of their chronological discovery in any one year.

find. Many have named them after their favorite house pets. Others have named them after spouses and sweethearts.

Meteors and Meteorites

While no asteroid has hit the Earth in recorded history, we are nevertheless continually being bombarded from outer space by cosmic rocks and dust particles, sometimes amounting to hundreds of tons every day.

Scattered between the planets, and in orbit round the Sun, are hordes of "micro-asteroids" and swarms of dust grains. When these cross over the Earth's orbit, they are drawn in towards us and enter our atmosphere at high speeds.

As a consequence of their high speeds, they begin to glow like the heat shield of a space craft on reentry. Most of the dust grains burn up completely, and if it is nighttime, they are seen as shooting stars, or *meteors,* as they are more correctly called. Some of the smaller rocks also burn up and these are seen as fireballs—a kind of very bright meteor which can be seen in broad daylight.

Larger space rocks are also seen falling during the daytime. If one is large enough, it will reach the surface of the Earth before it burns up completely. When they reach the ground, these objects are called *meteorites.*

Some meteorites are so large they were probably mini-asteroids before they hit the Earth. Several such objects have fallen in the past. One gouged out a huge crater in Arizona. It measures four-fifths of a mile in width and is 550 feet deep. Meteor Crater, as it is called, provides a popular tourist attraction, and you can walk right down into the bottom of the crater.

However, most cosmic rocks which reach us are smaller objects, and hundreds fall every year. Local and national museums hold large collections of the smaller examples. Some meteorites contain

Meteor Crater is located near Flagstaff in the northern Arizona desert.

A Russian stamp commemorating the fall of several large meteorites in eastern Siberia on February 12, 1947. As the meteorites descended, they left a huge trail of thick dust in their wake, as illustrated.

large amounts of nickel-iron, and some are pure nickel-iron. One of the most famous iron meteorites, which weighs 36½ tons, is displayed in the Museum of Natural History in New York.

Meteorites, as falling bodies, are not especially dangerous to the public at large. There is no genuine record of one having killed a person, but a lady in Alabama had her arm broken by one some years ago.

Watching Shooting Stars

Watching for shooting stars, or meteors as we shall call them from now on, is a naked-eye project. It needs no particular skill only a sharp eye and lots of patience.

Whereas nearly all cosmic rocks probably come from the asteroid belt, cosmic dust grains come from comets.

Comets are bodies that revolve around the Sun in egg-shaped, elliptical orbits. During the course of their travels, they often pass inside the Earth's orbit. In doing so they leave behind them a wake of dusty material. It is when the Earth encounters dense concentrations of this dust that we see brilliant and spectacular meteor showers, but these are rare events. Most of the meteors you see on a clear, moonless night originate from thinly spread-out dust particles plunging through the Earth's atmosphere. As a consequence you will count no more than about six meteors per hour on average. However, during the rare meteor spectaculars—often called meteor storms—you may see hundreds or even thousands flash across the sky every hour!

One famous meteor shower that occasionally produces a storm is called the Leonid meteor shower, or simply the Leonids. It gets this name because the meteors *appear* to diverge from the constellation

The Hunt for Cosmic Rocks

Hunting for cosmic rocks can be as exciting and rewarding as prospecting for gold. I know from experience because I have done and found both.

Meteorite falls in any one area are fairly rare events. You have to keep your eyes and ears alert for press, TV, or radio reports about falls occurring locally. Quite often, on its way down to Earth, a meteorite breaks up into smaller pieces, and these fragments spread out over a wide area. As a result some pieces get overlooked by the first searchers and are found weeks, months, or years later by those who have learned to recognize the telltale signs of a cosmic "nugget."

Cosmic "nuggets" turn up in the most unlikely places. When a meteorite fell at Barwell in England on Christmas Eve 1965, I later visited the fall area to prospect for fragments that had been overlooked. At first my companion and I had no luck. Then, following up a hunch, we asked permission from a local factory owner to search the false ceiling of his premises.

No sooner had we entered the false ceiling than we saw a patch of daylight above! There, on the boarded floor below, were several fragments of the meteorite. These had lain undisturbed for several weeks, hidden from view while searchers had scoured the countryside round about.

I still keep one of the fragments on my desk as a lucky cosmic charm. Before it was deflected towards the Earth, it was once part of a mini-asteroid in orbit around the Sun. As a skywatcher's very special curio, I would not part with it for all the tea in China.

of Leo. It must be understood, however, that the background stars of Leo are many light years away, while the meteors occur in the Earth's atmosphere less than 100 miles up. The Leonids occur every year in November. There were big displays in 1799, 1833, 1866, 1899, and 1966. There is a strong possibility another will occur in 1999, and skywatchers around the world will be on the lookout.

There are about twenty or so regular major showers which take place at definite dates throughout the year. All of these, like the Leonids, get their names from the constellations they appear to diverge from (see below).

A contemporary artist's impression of the Leonid Meteor Storm as seen in 1799 from South America by the famous travellers and scientists Humboldt and Bonplan.

The Major Annual Meteor Showers

Shower	Date of Peak Activity	Radiant Coordinates RA	DEC	Duration of Detectable Meteors	Duration of Peak Days	Expected Hourly Rates
Quadrantids*	Jan. 3	231°	+50°	Jan. 1–4	0.5	50
Corona Australids	Mar. 16	245	–48	Mar. 14–18	5	5
Virginids	Mar. 20	190	00	Mar. 5–Apr. 2	20	5
Lyrids	Apr. 21	272	+32	Apr. 19–24	2	10
Eta Aquarids	May 4	336	00	Apr. 21–May 12	10	20
Ophuichids	June 20	260	–20	June 17–26	10	20
Capricornids	July 25	315	–15	July 10–Aug. 5	20	20
Southern Delta Aquarids	July 29	339	–17	July 21–Aug. 15	15	20
Northern Delta Aquarids	July 29	339	00	July 15–Aug. 18	20	10
Pisces Australids	July 30	340	–30	July 15–Aug. 20	20	20
Perseids	Aug. 12	46	+58	July 25–Aug. 17	5	50
Kappa Cygnids	Aug. 20	290	+55	Aug. 18–22	3	5
Orionids	Oct. 21	95	+15	Oct. 18–26	5	20
Southern Taurids	Nov. 1	52	+14	Sept. 15–Dec. 15	45	5
Northern Taurids	Nov. 1	54	+21	Oct. 15–Dec. 1	30	5
Leonids	Nov. 17	152	+22	Nov. 14–20	4	varies
Phoenicids	Dec. 5	15	–55	Dec. 5	0.5	50
Geminids	Dec. 13	113	+32	Dec. 7–15	6	50
Ursids	Dec. 22	217	+80	Dec. 17–24	2	5

* This name refers to the obsolete constellation of Quadrans Muralis, now part of Draco.

One of the fascinating things about meteor watching is that no one can be certain when the next big new shower or storm will occur. The unexpected can always happen, and if you are outdoors and looking at the sky, you may be treated to the sight of a lifetime.

Meteor-watch parties can be fun social events! Why not organize one locally on a night when an annual shower is due to display?

Comets

Discover a new comet and you earn the privilege of calling it by your family name.

Maybe you think the discovery of a new comet sounds a little too ambitious for a young skywatcher. Take my word for it, by diligent searching *anything* is possible. If you are lucky, you may find a new comet very early on in your skywatching career.

This happened to American schoolboy Mark A. Whitaker while observing with a 4-inch reflecting telescope equipped with a x45 eyepiece. In 1969 he discovered a new comet on his third night outdoors! This comet has since been known in the record books as Whitaker's comet.

However, Mark did have a certain amount of beginner's luck. Another skywatcher, a good friend of mine, spent many years searching for comets without any luck at all. Then, to his surprise, he discovered two new comets within a week and, over the next few years, three more. Lady Luck had finally noticed him.

Most of us have to be content with looking at comets discovered by other people. Comet watching can be fascinating because usually half a dozen or more comets are within range of small telescopes, and some of them can be seen with binoculars.

Spectacular comets, with long tails that stretch right across the sky seen illustrating the pages of popular astronomy books, are fairly rare events, and their timing of appearance is quite uncertain. Some years, simply by chance, two or more bright comets become prominent naked-eye objects, then follows a long gap lasting several years without any. It is this factor of never knowing when the next bright comet will come along which makes comet watching so fascinating.

Comets themselves are curious nebulous bodies and, like the planets, revolve in orbits around the Sun. However, they differ from the planets in that many comets have very elongated orbits that are also often steeply inclined. Therefore they are often seen in parts of the sky not visited by the planets.

One special family of comets, called sungrazers, are sometimes so bright they can be seen in daylight. They get their name from the fact that they approach to within a few thousand miles of the Sun's surface before they pass around it and dash headlong back into deep space. Some of these comets break up into fragments and then visit the Sun again, hundreds or even thousands of years later, as separate comets.

Comet tails are made up of dust and gas but are so flimsy the Earth can pass right through one without our noticing any effects. This happened when Halley's famous comet revisited the Sun in 1910 and the Earth passed right through its broad tail.

Until Halley's comet last revisited the Sun in 1985/86, after its 76-year journey to beyond Neptune and back, little was known about the structure inside a comet's head. In 1986 Halley's comet was photographed at very close range by the *Giotto* and *Vega* space probes. Inside the nebulous head a solid body, or nucleus, revealed

Halley's famous comet returned to the Sun in 1986 after its 76-year journey out beyond the orbit of Neptune and back.

itself. This was an elongated mass made up of dust and ices measuring 9 miles long by 5 miles wide, and jets of gas were seen spouting from it. While astronomers had long suspected that a comet's head was probably like a large "dirty snowball," this was the first time one was actually photographed.

Beginners to skywatching are sometimes confused about how swiftly a comet moves across the sky. Because bright comets have long tails, they give the impression that the comets are moving as fast in our atmosphere as shooting stars. Newspapers and magazines are often to blame for this confusion. Journalists frequently write: "It shot across the sky as swiftly as a comet." While a shooting star does move swiftly and is seen as a flash of light in the atmosphere, a comet, because it is much farther away from us, appears to move more slowly. Its shift across the starry background is only seen from hour to hour and from night to night.

Comets are probably the leftovers from when the planets were formed in the solar system. It is believed that beyond the orbit of Pluto there lies a vast unseen family of comets. The ones we do see are only those that become disturbed in their orbits and are pulled in towards the Sun. These comets visit us periodically and have paths that carry them around the Sun; these range from as little as just over three years, to, in the case of some of the sungrazers, fifty or more thousand years. Halley's comet is a typical member of the periodic comets. It has been seen regularly every 75 to 76 years for over two thousand years.

Comet Hunting

Because comets brighten up when they are nearest the Sun, many comet hunters choose to search for them in the western sky after sunset and in the eastern sky before dawn. Nevertheless, new comets can be found *anywhere* in the sky—even in daylight hours.

Some new comets are found by accident. A new comet may wander into the neighborhood of a variable star, and its discovery falls to a skywatcher who happens to be looking at this star at the right time. Such discoveries have often happened in the past, and we can be sure they will happen again in the future, so be on guard!

You can even comet hunt with the naked eye, and several comets have been found this way in recent times. But the best chances of discovery are with small telescopes or binoculars. Prefer those which give a wide field of view. With small telescopes, a minimum magnification of about x25 is best because small comets are often overlooked with instruments magnifying less.

Binoculars make ideal instruments for comet viewing and hunting. Especially good are ex–World War II military-surplus binoculars. Some of these from Europe have lenses 4 inches in diameter and magnifications of x20 or x25. In binocular jargon these are called 20x105s or 25x105s. With these kinds of binoculars many comets have been discovered by amateur skywatchers.

Large 25 × 105 ex-military surplus binoculars. These are especially useful for galaxy and cluster spotting and for comet hunting.

Even with ordinary binoculars several comets have been found. In Argentina an amateur skywatcher and priest, Frederic William Gerber, found two comets with 8x24s (whose front lenses measured less than an inch across). He later discovered another comet using 12x60 binoculars.

Bright comets often appear in the dawn sky and are seen by airline pilots flying long-distance routes. One bright comet was first spotted by an Israeli air hostess when she looked through the cabin window. The tail of another bright comet was spotted by the caretaker at the Mount Wilson Observatory, but he thought it was a searchlight and failed to report it. Days later, when the whole comet moved into the morning sky, he realized what he had been looking at earlier, but it was too late. By then, the comet had already been claimed by someone else.

Tailless comets are less easy to recognize. In appearance they look like regular nebulae or hazy star clusters. When coming across a fuzzy object in a sky sweep, you must first check it out on a star map to be sure you are not looking at a nebula or cluster. If it looks as if it *may* be a comet, you then have to watch it to see if it moves in relation to the background stars. If the object is a comet, it will show a slight movement over one hour.

The Comet Hot Line

If you do find a suspicious fuzzy object which you think may be a comet, *do not ring the newspapers, yet!* It may well be you have found one, but chances are it is a comet that astronomers already know about.

Because both new and regular periodic comets are coming and going continually, in a never-ending procession, it is impossible to say what comets will be visible in *your* sky when you read this book. To check out an object you suspect might be a comet, phone an astronomer at your local professional observatory or, failing that, contact a knowledgeable amateur skywatcher. Both will be pleased to help you, and they will have at hand up-to-the-minute information about what comets are visible. If you have discovered a comet, they will help you claim it.

Finally, when you are a little weary of searching the skies, remember that the arrival of any new comet is unpredictable. There is always a chance one will appear tonight!

Unidentified "Flying" Objects

Who has not heard of UFOs, and who is not fascinated by the thought of spotting an alien spacecraft?

Alas, all those known about for sure have been conjured up by imaginative writers and film directors like Steven Spielberg. That is not to say that one day an alien craft *may not* be spotted. In science you have to keep an open mind.

As a skywatcher, I promise you, you will see plenty of UFOs. These unidentified flying objects will greatly puzzle you. Some will only puzzle you for a minute or two. Then in a flash you will realize what they are. But other UFOs will keep you searching for possible explanations for years after.

Most UFO reports are the result of ignorance or lack of knowledge on the part of the eyewitness. Laypersons often report Venus, Mars, or Jupiter as UFOs when these planets are at their brightest. With the Planet Tables in the next chapter at hand, these particular "UFOs" should not lead you astray for long.

It is not surprising that a falling meteorite or an exploding fireball often gives rise to local UFO reports. For eyewitnesses these can be very scary events—with lots of flashing lights, loud bangs, and very strange sounds like express trains rushing headlong towards you. Other sounds heard at this time are more spooky—like the wind "singing" through the telegraph wires. Very peculiar, softer noises travel ahead of a falling meteorite and may cause horses and cattle to panic and then stampede.

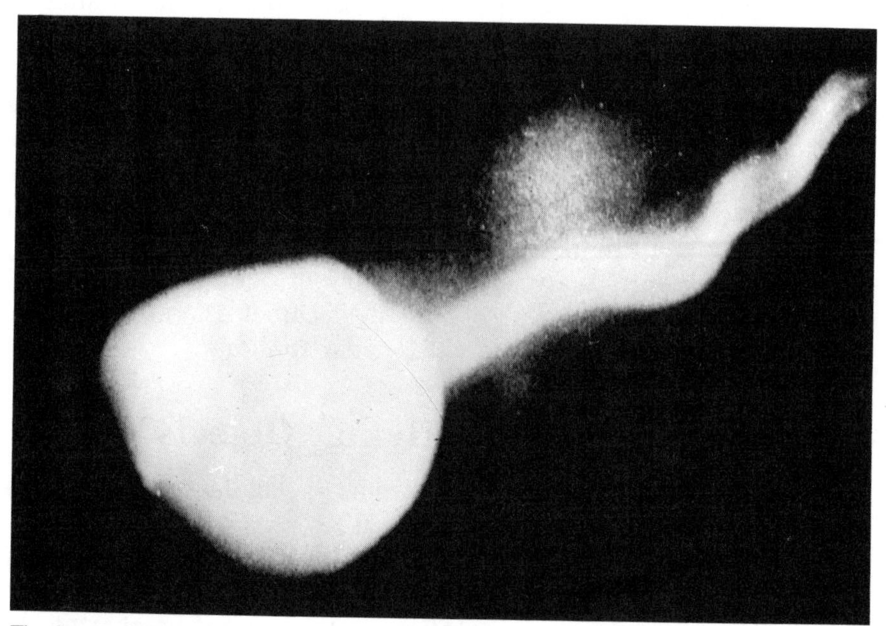

The famous March 24, 1933, "UFO" photographed over New Mexico. It later proved to be an exploding meteorite which scattered its cosmic debris along a 28-mile strip.

One man, who saw the Barwell Meteorite fall of 1965, told me it was exactly like a dive-bombing attack he had survived during World War II. One fragment fell so close it broke his front parlor window and then bounced into a flower vase.

Earth-orbiting man-made satellites also give rise to many UFO reports. There are now hundreds in orbit at any one time, plus all the debris from the launch vehicles that put them into space. Some of them appear to flash out messages in dots and dashes of the Morse code. To the more imaginative watcher this suggests an alien craft signalling to land. Not quite, however. The dots and dashes are more likely caused by an old rocket casing tumbling helter-skelter through the atmosphere, sometimes reflecting more sunlight sometimes less.

There are many dedicated skywatchers who specialize in looking at artificial satellites, and several organizations are devoted to their study. Addresses in astronomical yearbooks will put you in touch.

One UFO that puzzled many very experienced skywatchers for a while in 1968 was the *Apollo 8* spacecraft as it orbited the Earth before taking off for the Moon.

In the twilight sky over southern England it appeared like a peculiar, small comet. My telephone did not stop ringing for hours as observers rang to ask if *I* had seen the "UFO."

In this instance it did not take us too long to figure out what it was. All the same, it did take us by surprise, so you too beware. There is a straightforward explanation to account for 99.9 percent of all UFO sightings. The remaining 0.1 percent? Well, who knows?

The Zodiacal Light

As the name suggests, the zodiacal light occurs in the same part of the sky as the zodiacal stars (the ecliptic or path in the heavens followed by the major planets).

Observing the zodiacal light is a naked-eye project. Neither binoculars nor telescopes will help you to see it more clearly.

The light we see along the zodiac takes the shape of a very diffuse cone. Usually it is more a gentle pearly glow which is widest in the direction where the Sun has set in the evening sky and widest in the morning sky where the Sun will rise. It then gradually narrows off as it merges with the dark middle background of the night sky.

Beware of confusing the zodiacal light with the red afterglow you often see shortly after sunset. The sunset afterglow is caused by fine dust particles suspended in the Earth's atmosphere, but the zodiacal light is caused by dusty interplanetary material located much farther away.

Much of this interplanetary dust is material shed from comets as they orbit the Sun. Comets lose a lot more material when they are closer to the Sun than they do when they are farther away. This material collects in the region of the inner planets. Sunlight is reflected off the fine dust and, at the distance we view it from the Earth, it appears like a gentle glow.

Unfortunately, it requires a very dark, moonless night to see the zodiacal light. It is best seen in semi-tropical and tropical countries where the ecliptic is more upright over the horizon. In these regions it can be bright enough in the early morning to look like the oncoming of daylight—hence it is often nicknamed "the False Dawn."

The Aurora—the Northern and Southern Lights

Auroral displays in the night sky are better known as the Northern and Southern Lights. These displays take place in the Earth's upper atmosphere and are triggered by events on the Sun 93 million miles away.

The cause of the Northern and Southern Lights was long a mystery to astronomers. Then it was discovered that the Sun emits streams of radioactive particles, particularly when sunspots are more active. These particles "blow" outwards from the Sun and travel towards the inner planets, forming what astronomers call the solar wind.

As the solar wind particles pass near the Earth, they are attracted by our north and south magnetic poles. This, in turn, triggers off some very complex electrical activity in our atmosphere. It makes the sky glow in different colors with strange, dancing patterns. When these displays occur in the northern hemisphere, they are called the Northern Lights or, by their scientific name, the aurora borealis (Latin for "dawn" and "in the north"); those occurring in the southern hemisphere are called the Southern Lights or the aurora australis (Latin for "dawn" and "in the south").

Like the zodiacal light, the aurora is a naked-eye project, and binoculars and telescopes can be left indoors.

Best chances of seeing the aurora are if you live in the northern regions of the United States, in Canada, and in the Scandinavian countries. In the southern hemisphere it is more often seen in New Zealand than in Australia. Nevertheless, around the time of sunspot maximum the aurora can be seen almost anywhere in the world. At these times, watch for reports in local newspapers, since the aurora is often active over several nights.

Once seen, the aurora is never forgotten. Sometimes the whole sky becomes alive with fantastic color patterns that change shape from moment to moment and leave you breathless with delight. I once took part in an expedition to the Antarctic, and one of my tasks was to record displays of the aurora australis from night to night. During my year in the south I lost a lot of sleep, but those dancing colored lights will live forever in my memory.

Finding the Planets

Using the numbers (1 to 359) given in the simple Planet Tables below, you can find in what constellation the Sun, Mercury, Venus, Mars, Jupiter, and Saturn are located during the next few years.

To locate the brighter asteroids, Uranus, and Neptune, see below.

Because we are working in numbers of celestial longitude (CL), we first have to mark up the Star Maps 2, 3, and 4 with sets of numbers. These are the same kind of numbers you find in the Planet Tables.

All you have to remember is that 1 hour of RA equals 15° CL.

Below, you will find the number equivalents of every hour of RA printed along the top edges of Star Maps 2, 3, and 4.

RA	=	CL	RA	=	CL	RA	=	CL	RA	=	CL
1	=	15	7	=	105	13	=	195	19	=	285
2	=	30	8	=	120	14	=	210	20	=	300
3	=	45	9	=	135	15	=	225	21	=	315
4	=	60	10	=	150	16	=	240	22	=	330
5	=	75	11	=	165	17	=	255	23	=	345
6	=	90	12	=	180	18	=	270	24/0	=	360/0

I suggest you pencil in these CL numbers alongside the existing RA hours on photocopies of each of the Star Maps 2, 3, and 4 (see example illustrated).

Now we are ready to start.

The Planet Tables can be used in two ways:

You can use them *to find* one of the brighter planets; or you can use them *to identify* an object you have spotted in the night sky which you think may be one of the brighter planets.

Identifying a Planet

Example 1 (illustrated): To locate Jupiter on October 18, 1998, note that the date falls between entries on October 13 and October 23. Select the appropriate star map (Number 4) then lightly mark in pencil the estimated CL (349½) along the top of the map. Now extend a vertical line downwards *until it intercepts the line of the ecliptic.* Near the ecliptic line, you will find Jupiter hovering on the borders of Pisces and Aquarius.

Example 2 (not illustrated): On November 4, 1996, you spy a brightish, unidentified object near the line of the ecliptic, which is not marked on Star Map 2 in a position around 0° CL (in the constellation of Pisces). By checking the Planet Tables around this date, you will identify the object as Saturn.

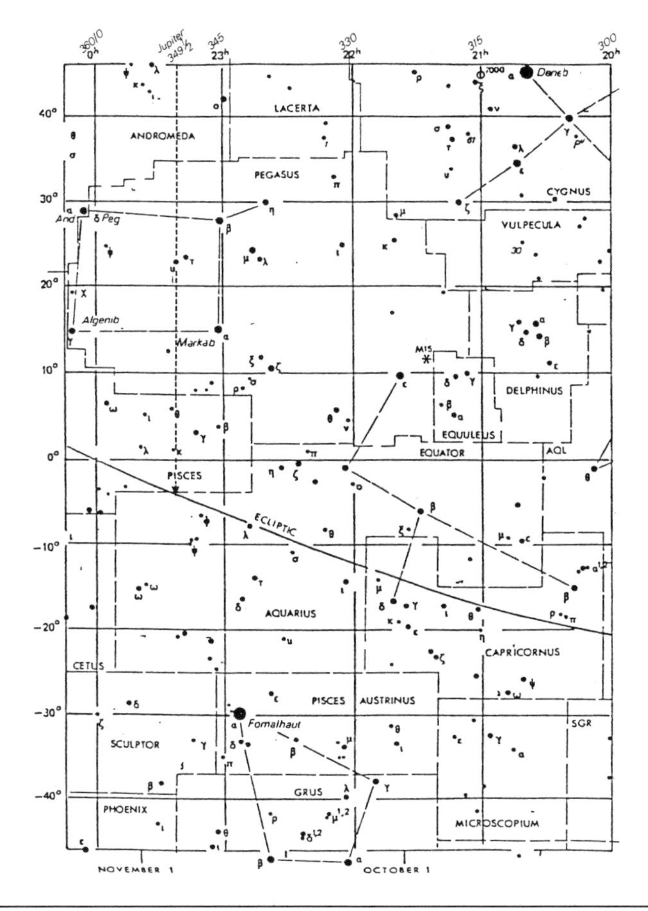

The tables are set for dates at 10-day intervals. The location of each of the brighter planets for those dates is given by a number ranging from 1 to 359 (the planet's celestial longitude).

To find a planet, look up its number in the tables for the date you are observing. If this date falls between two 10-day intervals, estimate the number you require using the number above and the one below.

Having noted the number, turn to the star maps and look for the map carrying the number you want.

Lightly pencil this in. Now drop a vertical line straight down until it intercepts the curved line on the star map labelled the *ecliptic* (see example). The intercept position marks the approximate spot where the planet is to be found.

Alternatively, if you wish to identify an object you think may be a planet, you work backwards, from star map to tables.

Lightly pencil in the position of the object on the star map. Then carry a vertical line straight upwards to intercept the line of numbers at the top.

Now estimate this number.

Having noted the number, look in the Planet Tables under the date you are observing and try to match it up.

You will not often find an *exact* match, but if the object you have spotted is one of the brighter planets, you should be close enough to make a positive identification.

A more direct method of planet finding is to use the CL number to find which constellation a planet is crossing. Using the list below, you can quickly scan the different numbers that lie in each of the constellations and then go direct to the star maps.

CL	Constellation	CL	Constellation
0–26	Pisces	215–239	Libra
26–50	Aries	239–245	Scorpius
50–89	Taurus	245–265	Ophiuchus
89–119	Gemini	265–301	Sagittarius
119–140	Cancer	301–329	Capricornus
140–174	Leo	329–351	Aquarius
174–215	Virgo	351–0	Pisces

Finding the Brighter Asteroids, Uranus, and Neptune

Even as a beginner you too can play sky detective and hunt for the brighter asteroids, Uranus, and Neptune.

However, up-to-date information about the location of these objects is outside the scope of this book. To glean the necessary information, you will need to visit your local library and seek the help of your librarian to obtain access to the publications mentioned below.

Asteroids

While one or two of the brighter asteroids just reach the threshold of naked-eye visibility, some kind of optical aid is required to identify them properly against the fainter background stars.

With 8x30 binoculars at least 11 asteroids come within range. With a small telescope scores are visible.

To locate an asteroid you first need to know which ones are currently visible in your sky. To obtain this information you will need to consult a current astronomical yearbook or, perhaps easier, one of the popular astronomical magazines, of which there are several. Some are available in public libraries.

The yearbooks and current magazines provide up-to-the-minute information and often include Finder Charts for particular asteroids. Otherwise you will have to plot the asteroid's position on a tracing from an advanced star atlas which shows stars well beyond naked-eye visibility. There are several star atlases available showing stars as faint as mag 9, or fainter.

Uranus and Neptune

The planets Uranus and Neptune can be found with binoculars in the same way as the brighter asteroids, using the same sources of information as above.

Uranus, shining at mag 6, is the easier object to spot. Neptune, at mag 7.6, needs a dark sky but is quite easy with a 2-inch telescope when you know *exactly* where to look.

To be sure you have spotted any planet correctly, draw your own star map. If it is a planet, you will see that it moves slightly from night to night against the background of fixed stars.

Planet-Finder Tables

Date			Sun	Mer	Ven	Mar	Jup	Sat
1993	Jun	11	80	104	34	143	185	330
1993	Jun	21	90	114	44	148	185	330
1993	Jul	1	99	118	55	154	186	330
1993	Jul	11	109	115	65	160	187	330
1993	Jul	21	118	109	76	166	188	329
1993	Jul	31	128	109	88	172	190	328
1993	Aug	10	137	120	99	179	191	328
1993	Aug	20	147	137	111	185	193	327
1993	Aug	30	157	157	123	191	195	326
1993	Sep	9	166	176	135	198	197	325
1993	Sep	19	176	192	147	205	199	325
1993	Sep	29	186	207	159	211	201	324
1993	Oct	9	196	220	171	218	203	324
1993	Oct	19	206	230	185	225	205	324
1993	Oct	29	216	232	196	232	207	324
1993	Nov	8	226	221	209	239	209	324
1993	Nov	18	236	217	221	246	212	324
1993	Nov	28	246	227	234	254	214	324
1993	Dec	8	256	241	246	261	216	325
1993	Dec	18	266	257	259	268	217	326
1993	Dec	28	276	272	271	276	219	327
1994	Jan	7	286	288	284	284	221	328
1994	Jan	17	297	305	297	291	222	329
1994	Jan	27	307	322	309	299	223	330
1994	Feb	6	317	335	322	307	224	331
1994	Feb	16	327	336	334	315	224	332
1994	Feb	26	337	326	347	323	225	333
1994	Mar	8	347	323	359	330	225	334
1994	Mar	18	357	330	12	338	224	336
1994	Mar	28	7	341	24	346	224	337
1994	Apr	7	17	356	36	354	223	338
1994	Apr	17	27	13	49	2	221	339
1994	Apr	27	37	33	61	9	220	340
1994	May	7	46	54	73	17	219	341
1994	May	17	56	74	85	25	218	341
1993	May	27	65	88	97	32	217	342
1994	Jun	6	75	97	109	40	216	342
1994	Jun	16	85	98	121	47	215	342
1994	Jun	26	94	93	132	54	215	342
1994	Jul	6	104	89	144	61	215	342
1994	Jul	16	113	93	155	68	215	342
1994	Jul	26	123	105	166	75	216	342
1994	Aug	5	132	124	177	82	216	341
1994	Aug	15	142	144	188	89	217	340
1994	Aug	25	152	163	198	95	219	340
1994	Sep	4	161	180	207	102	220	339
1994	Sep	14	171	194	215	108	222	338
1994	Sep	24	181	207	222	114	224	337
1994	Oct	4	191	215	226	120	226	337
1994	Oct	14	200	215	228	125	228	336
1994	Oct	24	210	204	226	130	230	336
1994	Nov	3	220	202	220	135	232	336
1994	Nov	13	230	213	215	140	234	336
1994	Nov	23	240	229	212	144	236	336
1994	Dec	3	251	244	214	147	239	336
1994	Dec	13	261	260	219	150	241	337
1994	Dec	23	271	276	226	152	243	337
1995	Jan	2	281	292	235	153	245	338
1995	Jan	12	291	308	244	152	247	339
1995	Jan	22	301	320	255	150	249	340

Date			Sun	Mer	Ven	Mar	Jup	Sat
1995	Feb	1	312	318	266	147	250	341
1995	Feb	11	322	327	277	143	252	342
1995	Feb	21	332	327	288	139	253	343
1995	Mar	3	342	315	300	136	254	345
1995	Mar	13	352	327	312	134	255	346
1995	Mar	23	2	342	324	133	255	347
1995	Apr	2	12	359	336	134	255	348
1995	Apr	12	22	19	348	135	255	349
1995	Apr	22	31	40	0	137	255	350
1995	May	2	41	59	12	141	254	351
1995	May	12	51	72	24	144	253	352
1995	May	22	60	78	36	148	252	353
1995	Jun	1	70	76	48	153	251	354
1995	Jun	11	80	71	60	158	249	354
1995	Jun	21	89	70	73	163	248	355
1995	Jul	1	99	77	85	168	247	355
1995	Jul	11	108	90	97	174	246	355
1995	Jul	21	118	110	109	180	246	355
1995	Jul	31	127	131	122	186	246	354
1995	Aug	10	137	150	134	192	246	354
1995	Aug	20	147	167	146	198	246	353
1995	Aug	30	156	181	159	205	247	353
1995	Sep	9	166	193	171	211	248	352
1995	Sep	19	176	200	183	218	249	351
1995	Sep	29	185	198	196	225	250	350
1995	Oct	9	195	187	208	232	252	350
1995	Oct	19	205	187	221	239	254	349
1995	Oct	29	215	200	233	246	255	348
1995	Nov	8	225	216	246	253	257	348
1995	Nov	18	235	232	258	261	260	348
1995	Nov	28	245	248	271	268	262	348
1995	Dec	8	255	264	283	276	264	348
1995	Dec	18	266	279	295	283	266	349
1995	Dec	28	276	294	308	291	269	349
1996	Jan	7	286	304	320	299	271	350
1996	Jan	17	296	300	332	307	273	351
1996	Jan	27	306	290	344	315	275	352
1996	Feb	6	316	292	356	322	277	353
1996	Feb	16	327	301	8	330	279	354
1996	Feb	26	337	314	20	338	281	355
1996	Mar	7	347	329	31	346	283	356
1996	Mar	17	357	346	42	354	284	357
1996	Mar	27	7	5	52	2	285	359
1996	Apr	6	16	26	62	10	286	0
1996	Apr	16	26	44	71	17	287	1
1996	Apr	26	36	56	79	25	288	2
1996	May	6	46	58	85	32	288	3
1996	May	16	55	54	88	40	287	4
1996	May	26	65	50	88	47	287	5
1996	Jun	5	75	52	83	55	286	6
1996	Jun	15	84	61	77	62	285	6
1996	Jun	25	94	76	73	69	284	7
1996	Jul	5	103	96	72	76	283	7
1996	Jul	15	113	117	75	83	281	7
1996	Jul	25	122	137	80	89	280	7
1996	Aug	4	132	154	87	96	279	7
1996	Aug	14	141	168	96	103	278	7
1996	Aug	24	151	178	105	109	278	6
1996	Sep	3	161	183	115	115	278	6
1996	Sep	13	170	179	126	122	278	5

Date			Sun	Mer	Ven	Mar	Jup	Sat
1996	Sep	23	180	170	137	128	278	4
1996	Oct	3	190	172	148	134	279	3
1996	Oct	13	200	186	160	140	280	3
1996	Oct	23	210	203	172	145	281	2
1996	Nov	2	220	220	184	151	283	1
1996	Nov	12	230	236	196	157	285	1
1996	Nov	22	240	251	209	162	287	1
1996	Dec	2	250	266	221	167	289	1
1996	Dec	12	260	280	233	171	291	1
1996	Dec	22	270	289	246	176	293	1
1997	Jan	1	281	283	258	179	295	1
1997	Jan	11	291	273	271	182	298	2
1997	Jan	21	301	277	283	184	300	3
1997	Jan	31	311	287	296	186	302	4
1997	Feb	10	321	301	309	186	305	4
1997	Feb	20	331	316	321	185	307	6
1997	Mar	2	341	333	334	182	309	7
1997	Mar	12	351	352	346	179	311	8
1997	Mar	22	1	12	358	175	313	9
1997	Apr	1	11	29	11	171	315	10
1997	Apr	11	21	39	23	169	317	12
1997	Apr	21	31	38	36	167	318	13
1997	May	1	41	32	48	167	320	14
1997	May	11	50	30	60	168	321	15
1997	May	21	60	35	73	170	321	16
1997	May	31	70	46	85	173	322	17
1997	Jun	10	79	62	97	176	322	18
1997	Jun	20	89	82	109	180	322	19
1997	Jun	30	98	103	121	185	321	19
1997	Jul	10	108	123	134	190	321	20
1997	Jul	20	117	140	146	195	320	20
1997	Jul	30	127	154	158	201	318	20
1997	Aug	9	136	163	170	207	317	20
1997	Aug	19	146	166	182	213	316	20
1997	Aug	29	156	161	193	219	315	20
1997	Sep	8	165	153	205	226	314	19
1997	Sep	18	175	157	217	232	313	19
1997	Sep	28	185	172	228	239	312	18
1997	Oct	8	195	190	240	246	312	17
1997	Oct	18	205	208	251	254	312	16
1997	Oct	28	215	224	261	261	313	16
1997	Nov	7	225	239	272	268	314	15
1997	Nov	17	235	253	281	276	315	14
1997	Nov	27	245	266	290	284	316	14
1997	Dec	7	255	273	297	291	318	14
1997	Dec	17	265	266	302	299	319	14
1997	Dec	27	275	257	304	307	321	14
1998	Jan	6	285	262	302	315	323	14
1998	Jan	16	296	274	296	323	325	14
1998	Jan	26	306	288	291	330	328	15
1998	Feb	5	316	305	288	338	330	16
1998	Feb	15	326	320	290	346	333	17
1998	Feb	25	336	338	295	354	335	18
1998	Mar	7	346	357	302	2	337	19
1998	Mar	17	356	14	310	10	340	20
1998	Mar	27	6	21	319	17	342	21
1998	Apr	6	16	17	330	25	344	22
1998	Apr	16	26	11	340	32	346	24
1998	Apr	26	36	11	351	40	349	25
1998	May	6	45	19	2	47	350	26

Date			Sun	Mer	Ven	Mar	Jup	Sat
1998	May	16	55	31	14	54	352	27
1998	May	26	65	48	25	61	354	29
1998	Jun	5	74	68	37	68	355	30
1998	Jun	15	84	90	49	75	356	31
1998	Jun	25	93	110	61	82	357	31
1998	Jul	5	103	126	72	89	358	32
1998	Jul	15	112	139	84	96	358	33
1998	Jul	25	122	147	96	102	358	33
1998	Aug	4	131	148	109	109	358	34
1998	Aug	14	141	141	121	116	357	34
1998	Aug	24	151	136	133	122	356	34
1998	Sep	3	160	143	145	128	355	33
1998	Sep	13	170	159	158	135	353	33
1998	Sep	23	180	177	170	141	352	32
1998	Oct	3	190	195	183	147	351	32
1998	Oct	13	199	211	195	153	350	31
1998	Oct	23	209	227	208	159	349	30
1998	Nov	2	219	240	220	165	348	29
1998	Nov	12	229	252	233	171	348	29
1998	Nov	22	239	258	245	177	348	28
1998	Dec	2	250	249	258	183	349	27
1998	Dec	12	260	241	270	188	350	27
1998	Dec	22	270	248	283	193	351	27
1999	Jan	1	280	261	295	198	352	27
1999	Jan	11	290	276	308	203	354	27
1999	Jan	21	300	291	320	208	355	27
1999	Jan	31	311	308	333	212	357	28
1999	Feb	10	321	325	345	215	359	28
1999	Feb	20	331	343	358	218	2	29
1999	Mar	2	341	359	10	221	4	30
1999	Mar	12	351	4	22	222	6	31
1999	Mar	22	1	357	34	222	9	32
1999	Apr	1	11	351	46	221	11	33
1999	Apr	11	21	354	58	219	13	35
1999	Apr	21	30	3	70	215	16	36
1999	May	1	40	17	81	212	18	37
1999	May	11	50	34	93	208	20	38
1999	May	21	60	54	104	206	23	40
1999	May	31	69	76	114	205	25	41
1999	Jun	10	79	96	124	205	27	42
1999	Jun	20	88	112	133	206	29	43
1999	Jun	30	98	123	142	208	30	44
1999	Jul	10	107	129	149	212	32	45
1999	Jul	20	117	128	153	216	33	46
1999	Jul	30	126	121	155	220	34	46
1999	Aug	9	136	119	153	226	35	47
1999	Aug	19	146	128	148	231	35	47
1999	Aug	29	155	145	142	237	35	47
1999	Sep	8	165	162	139	243	35	47
1999	Sep	18	175	183	140	250	34	47
1999	Sep	28	184	199	144	257	33	46
1999	Oct	8	194	214	150	264	32	46
1999	Oct	18	204	227	158	271	31	45
1999	Oct	28	214	238	168	278	29	44
1999	Nov	7	224	241	178	285	28	44
1999	Nov	17	234	232	189	293	27	43
1999	Nov	27	244	226	200	301	26	42
1999	Dec	7	254	235	211	308	25	41
1999	Dec	17	265	248	223	316	25	41
1999	Dec	27	275	263	235	324	25	41

Index

Note: Figures in bold refer to illustration numbers in the color section; figures in italics refer to page numbers where a black-and-white text illustration appears.

About the Author

Peter Lancaster-Brown is a noted writer, broadcaster, and lecturer in the field of astronomy.

A Fellow of the Royal Astronomical Society, and a past President of the Junior Astronomical Society, he is also a member of the Comet Commission of the International Astronomical Union, the official world body responsible for monitoring the comings and goings of comets.

His many books include *Star and Planet Spotting*.